BEYOND THE KNOWN

BEYOND THE KNOWN

HOW EXPLORATION CREATED THE MODERN WORLD AND WILL TAKE US TO THE STARS

ANDREW RADER

SIMON &
SCHUSTER

London · New York · Sydney · Toronto · New Delhi

A CBS COMPANY

First published in the United States by Scribner,
an imprint of Simon & Schuster, Inc., 2019
First published in Great Britain by Simon & Schuster UK Ltd, 2019
A CBS COMPANY

1 3 5 7 9 10 8 6 4 2

Simon & Schuster UK Ltd
1st Floor
222 Gray's Inn Road
London WC1X 8HB

www.simonandschuster.co.uk
www.simonandschuster.com.au
www.simonandschuster.co.in

Simon & Schuster Australia, Sydney
Simon & Schuster India, New Delhi

A CIP catalogue record for this book is available from the British Library.

Hardback ISBN: 978-1-4711-8647-9
Trade Paperback ISBN: 978-1-4711-8648-6
eBook ISBN: 978-1-4711-8649-3

Printed and bound by CPI Group (UK) Ltd, Croydon, CR0 4YY

MIX
Paper from
responsible sources
FSC
www.fsc.org FSC® C020471

Only those who will risk going too far can possibly find out how far one can go.

—*T. S. Eliot*

CONTENTS

PART III: MODERNITY

PART IV: BECOMING *STAR TREK*

INTRODUCTION

On December 21, 2015, SpaceX landed a reusable rocket after successfully launching a satellite into orbit—the first time this feat had ever been accomplished. That Falcon 9 rocket now stands proudly outside company headquarters in Hawthorne, California. I pass it on my way into work.

Why does this event matter? Let me ask the question differently. What happened in 1492? In January 1492, the last Moorish kingdom of Spain surrendered, ending a 781-year Muslim occupation that had once threatened all of Europe. In March of that same year, Scotland and France renewed a two-century-old anti-English alliance. Later that month, Spain declared that it would expel all its Jews, displacing more than 100,000. In May, a riot in the Netherlands killed 232 people. In August, a rapacious candidate bribed his way to the papacy, and was beset by scandal when it was revealed that he'd fathered several children with multiple mistresses. In October, Henry VII of England concluded peace with the French after leading a cross-channel invasion of the continent. In November, a meteorite struck a wheat field in France, creating a fireball visible for a hundred miles.

These events shook the lives of millions, but what do we remember about 1492? Columbus sailed the ocean blue. Taking the long view of history, we see that five hundred years hence, petty political squabbles, celebrity gossip, and fluctuations of the stock market won't matter. But exploration will. Columbus was important not because of his individual achievements but because he launched a new age of discovery that connected people around the world. He pushed outward the borders of the "known."

So, too, did the engineers of SpaceX in December 2015—at least, in a first-step incremental way. The rocket landing was important because we live on a large rocky planet with intense gravity. It's really hard to get to space. Rockets have to fly at the technical edge of feasibility, carrying more than 90 percent of their total weight as fuel. To maximize performance, it's been standard practice to make rockets disposable. But using disposable rockets is like throwing away the airplane after each flight. Reusable rockets make space affordable, and this is one of the innovations that will open up the cosmos to human activity.

Ultimately, all exploration is an investment in our future. Most of the benefits of expanding into space will be realized by our future descendants, as has always been the case throughout history when people choose to look beyond their horizons. Asking why we should travel beyond Earth is like asking our early ancestors why they should leave the confines of the African Rift Valley. Most needs are being met, so why leave? But maybe there are new sources of food over the hills, or solutions to problems that can only be found by venturing into the unknown.

Crucially, by placing ourselves at the leading edge of what's possible, we create incentives to solve problems that haven't been solved before, often with unforeseen applications. Columbus sailed with flimsy coastal vessels unsuited to rough Atlantic waters because oceangoing sailing ships hadn't yet been invented—and never *would* have been without the awareness of new continents across the sea. Without an ocean to cross, we never would have invented passenger liners or transcontinental air transport. At the beginning of the Cold War, America didn't know how to send people to space, but in trying to figure it out, NASA invented life-support technologies, water filtration systems, cordless power tools, fireproof clothing, wireless data transfer, solar panels, insulin monitors, remote control systems, weather forecasting, medical scanning technologies, and more than two thousand other spin-offs.

This is a book about how exploration enriches us. It's a story of discovery and adventure, of wealth and conquest, of prejudice and tolerance.

The first section begins with the first wave of human expansion and then follows the voyages of the ancients, from the Polynesians to the Egyptians to the Greeks, and right up to the fall of Rome. As we shall see, these civilizations well understood that exploration, trade, and the exchange of ideas were critical to their prosperity.

The book's second section picks up after Rome's fall, starting with the Vikings, and brings us to Magellan's expedition around the world. It was during this "Age of Exploration" that most of our planet was connected to form our modern global system.

The book's third section takes us from the scientific voyages of discovery into the skies as humans masters flight, and through the space race up to modern times. With the world now more connected than ever, we might ask: Is there anything left to discover? The answer is most definitely *yes*. Recent data from planet-hunting missions such as the Kepler Space Telescope suggest that there are billions of Earthlike planets in our galaxy alone, among hundreds of billions of galaxies. So most of the story of exploration actually lies ahead of us, unwritten.

In the book's fourth section, I'll try to shed light on what future "beyond Earth" explorations might look like. Among the questions I'll attempt to answer: Why is Mars the most important goal in the near term, and how will we get there, live, and prosper? What's next *after* Mars? Will we ever be able to travel to another star? And what is the ultimate destination for humanity?

Whatever the answers to those questions, it's clear that our civilization is at a crossroads. For a hundred thousand years, we lived in small bands spread across the planet's surface, with no knowledge of anything beyond our immediate surroundings. It's only in the last few centuries that the human family has been reunited to share a common awareness. Together, we face tremendous challenges in the form of population growth, environmental depletion, and resource exhaustion, but our most valuable tools are the same as those that carried us to this point: curiosity, drive, collaborative problem solving, and imagination. I believe that a bright future awaits our species. We have plenty more exploring to do.

PART I

IN THE BEGINNING

1 | OUT OF THE CRADLE

Humans aren't native to all of planet Earth. We're descendants of a small group of primates who evolved for millions of years in the Great Rift Valley of East Africa. It's our technology that enabled us to expand across the globe, starting around a hundred thousand years ago. There were earlier waves of proto-human migration, to be sure. *Homo erectus* was the first to leave the cradle, beginning its migration across Eurasia around 1.5 million years ago. Our cousins in that hominid line were the first to reach the Middle East, China, and Southeast Asia, and the first to command fire. *Homo erectus* also hunted large animals for the first time, and developed sophisticated tools. They may have built rafts, and even crossed large bodies of water, but it seems that once they left the cradle, they never returned. Theirs is not our story.

Next to leave Africa, before eight hundred thousand years ago, was a group appropriately named *Homo antecessor*. These wanderers may have been progenitors not only of us, but also of our closely related cousins the Neanderthals (there is some debate about the precise lineage). *Homo antecessor* looked essentially like us but was on the stocky side with a slightly smaller skull and brain. The smaller brain conferred several advantages: it consumed far less energy than our copious gray matter, and it permitted faster development and maturation. Whereas modern humans can't reproduce until around age twelve, a typical *Homo antecessor* functioned as an adult by eight or nine.

Around six hundred thousand years ago, *Homo antecessor* gave rise to *Homo heidelbergensis*, probably the first species to develop a sophisticated culture with a fully vocalized language, a ritual of burying their dead (proto-religion?), and cave art (we've found the remains of dye but no paintings per se). With more advanced technology (tools, fire, animal-skin clothes), *Homo heidelbergensis* bands reached the cold climates of Europe and Siberia. When *Homo heidelbergensis* expanded through Eurasia, it would have encountered *Homo erectus* and its descendants, who were already living there. In fact, for most of history, there've been many species of humans on Earth. It's actually quite astonishing that we live at a time when there's only *one* type of human. This situation has only persisted for the last thirty thousand years or so—less than 1 percent of the time hominids have been around. Undoubtedly, there were many past interactions among different human species. What would those encounters have been like?

It's possible that different human species mostly ignored each other, or they might have cooperated and traded. At the other extreme, they might have hunted each other. Groups of chimpanzees are known for raiding, killing, and even eating other primates, including other chimpanzees of the same type. And there's some evidence for cannibalism among early hominids (as well as some modern humans, to be sure). Would different human species have regarded each other as "people" or as types of "animals"? Perhaps neither. Diverse animals often congregate around a water hole without showing hostility. Maybe early hominids regarded each other with little more than mild curiosity.

While these encounters between *Homo heidelbergensis* and *Homo erectus* were playing out across Eurasia, our immediate ancestors were still confined to a small area in East Africa. The next time someone asks you where you're from, you can say, "Somewhere near Lake Turkana." There weren't many of us, and we almost didn't make it. At times, our entire population dwindled to a few thousand individuals. One of the expansion bottlenecks may have been the eruption of the giant Toba supervolcano on the island of Sumatra about seventy-five thousand years ago, which spewed enough ash into the atmosphere to trigger a six-year volcanic winter. These bottlenecks are the reason that humans are among the least genetically diverse species on Earth. Despite the superficial variation in human appearance around the world, humans are exceedingly similar. We know this from gene studies that measure the divergence of mitochondrial DNA as human groups became separated over time.*

It's unclear where we should draw the arbitrary line marking modern humans. The oldest fossil indistinguishable from a modern human is around 195,000 years old, uncovered in Kenya in 1975. But perhaps a better indicator is culture. Blombos Cave in South Africa, dating back 100,000 years, represents one of the earliest sites that shows a full range of modern human behavior. Evidence there suggests diverse resource use, multistep and multimaterial tool construction, complex art, social organization, and ritualistic behavior. Specific finds include the remains of shellfish, birds, turtles, ostrich eggs, ochre dye, engraved bones, detailed stone tools, and seashell beads used for decoration. (The "modernity" of the cave's occupants lies in the variety of animals they ate and their sense of vanity.)

The migration of some members of our species out of Africa is sometimes called "Out of Africa II" to contrast it with previous hominid departures, but don't picture hordes of people on the continent's borders waiting for the starter's whistle to blow. The dates are imprecise, the numbers of migrants very small, and the areas covered enormous. There's also a good deal of disputed evidence. Most scientists think that there were actually two waves of *Homo sapiens* leaving Africa—the first wave leaving around 120,000 years ago but not getting much farther than the Middle East, the second decisive wave departing 50,000 years later. Either way, between 120,000 and 70,000 years ago small bands of modern humans left Africa, never to return. By 50,000 years ago at least some adventurers had reached Australia, either by means of a land bridge during a period of low sea level or by traveling on

---

* Unlike cellular DNA acquired from both parents, mitochondrial DNA is passed down only through the mother. Since it doesn't vary much from generation to generation, it provides an excellent record of historical lineage.

boats from Indonesia. By around 40,000 years ago, humans had spread into Europe and Siberia, some coming up from the Middle East, others crossing at Gibraltar into Spain. By around 14,000 years ago, humans had arrived in the Americas for the first time, by crossing from Siberia into Alaska.

Modern humans were the first people to reach Australia and the Americas, but everywhere else they would have encountered people from earlier waves of migration. In Europe and the Middle East our ancestors had significant and sustained contact with Neanderthals, who'd been living there for hundreds of thousands of years. Neanderthals weren't so different from us. They were stronger, and possibly smarter. They crafted sophisticated tools and probably spoke a verbal language. They built simple boats and hunted large animals like mammoths. Stocky, barrel-chested, and powerful, they were at least a match for us on an individual basis. Neanderthals were so closely related to *Homo sapiens* that interactions between the two were probably no different than among two tribes.

Recent evidence suggests that Neanderthals were already in decline by the time our ancestors arrived, but it seems likely that there were at least *some* bloody conflicts as tribes of early *Homo sapiens* moved in to occupy the same territories. One of the more interesting notions put forth is that our ancestors carried diseases that Neanderthals hadn't been previously exposed to, foreshadowing contact between the Old World and the Americas. However, the likely truth is that Neanderthals were simply outcompeted. Perhaps better technology enabled our ancestors to more efficiently hunt animals and gather food. This is especially plausible since Neanderthals' larger bodies and brains would have required more nourishment, making them more susceptible to starvation.

Whatever the reason, between around forty thousand and twenty-eight thousand years ago, Neanderthals disappeared from the planet—but not entirely. The Neanderthal Genome Project has confirmed that significant portions of Neanderthal DNA still exist in modern humans. The exact amount varies by geographic location (less in Africa, where modern humans never left), but many populations seem to have up to 4 percent Neanderthal-sourced

DNA.* This isn't enough to represent a complete convergence of humans and Neanderthals, but there was clearly some mixing going on. If your ancestors are from outside of Africa, you may want to blame your Neanderthal heritage for your bad knees or inability to stick to your diet—though, of course, it won't do you much good.

Following *Homo sapiens*'s exodus from Africa, our species spread out to occupy every corner of the planet, with the exception of Antarctica and a collection of islands far out at sea. Along the way, we perfected the command of fire, the use of animal skins for clothing and shelter, and the manufacture of stone and bone tools. We domesticated dogs to help us hunt and guard our camps, or perhaps they domesticated themselves by self-selecting for tameness.† We invented the elements of culture, with carvings, cave paintings, and objects for ritual burial. Around twenty-five thousand years ago we invented pottery, ropes, harpoons, saws, sewing needles, braided baby carriers, baskets, and fishing nets. Also around this time, the first permanent settlements appeared, such as the ancient rock-and-mammoth-bone village unearthed at Dolní Věstonice in the Czech Republic.

What really changed the way humans lived on a massive scale was agriculture. It all began when hunter-gatherers in the Middle East, China, and Mesoamerica began sticking around longer in places that supported naturally growing wild grasses—the ancestors of wheat, barley, oats, rice, corn, and every other grain we know today. It just made sense to select the best of these wild grains to preferentially plant for harvesting the next year. Thus, totally inadvertently, these early people set humanity on a course

---

* There are examples of mixing with *other* types of humans. Around 3 to 5 percent of the DNA in Southeast Asian populations seems to come from Denisovans, a group that diverged from our ancestors slightly before Neanderthals.

† Only the boldest but most tolerant wolves would approach human camps in search of scraps and then actually stick around long enough to find out if they could manage to live with such strange two-legged creatures. These tamer wolves would be more successful because they could access human-sourced food. Essentially, dogs are good-natured wolves who like to eat garbage. So don't get upset at your dog's rummaging through the trash can in the middle of the night. It's in their nature.

toward sedentary living. Was this a good or bad thing? On an individual level, some hunter-gatherers managed healthier and more varied diets, but only in times of plenty when they could find enough food. Many more people can be consistently supported on grain. Today over half the calories consumed by humans on Earth come from just three types of grass: corn, rice, and wheat. Agriculture has been the essential ingredient of civilization, allowing us to support large populations that can work together.

For 97 percent of our existence, humanity had lived in small nomadic bands of fewer than a hundred individuals, organized around extended families. Living together with strangers required the development of power structures and laws to govern the behavior of a species still prone to restless violence. Communal interaction accelerated technological innovation through specialized production, collaborative problem solving, and the ability to exchange ideas. We domesticated animals on a large scale and learned to forge metals. We developed economies, religions, and armies in financial, cultural, and military struggles with neighbors. Agriculture put an end to our nomadic wandering but established the conditions for massive technological expansion, which would give us the tools to once again expand outward in a second wave of exploration.

2 | EARLY WANDERINGS

Unlike earlier hominids, *Homo sapiens* were the first to venture into Australia and the Americas. Australia came first, around fifty thousand years ago. With more water locked up in polar ice, sea levels would have been lower. Although the three hundred miles separating Indonesia from Australia is much too far to see across even on a clear day, more land would have been revealed by a lower sea level. This route still would have required crossing a hundred miles of open water, but this is possible in even the simple wooden rafts and dugout canoes built by Aboriginal Australians before European contact. Given thousands of years, it seems inevitable that small groups of humans would cross the Timor Sea to Australia, either intentionally or unintentionally. Indeed, refugees from Indonesia still cross by this route using primitive rafts.

Australians diverged into different communities, each specializing in local resources. Australia is a dry but varied continent ranging from tropical rain forest to temperate grassland to frosty mountains to deep desert. Most Aborigines remained nomadic, digging for edible roots and collecting fruits, berries, seeds, and insects. They hunted lizards, bandicoots, birds, possums, and snakes with spears. Kangaroos and other large animals were often disabled by a thrown club or boomerang. Aboriginal hunters are excellent trackers and stalkers, able to approach their quarry using cover and staying downwind or masking their smell. They learned to use animal pelts as disguises and mimicry to draw in inquisitive animals for an ambush. They used fire to expand hunting grounds or encourage the growth of particular plants. They captured fish by hand, stirring up the muddy bottom to chase them out or sprinkling the crushed leaves of poisonous plants in the water to paralyze them.

The development of fishing technologies such as spears, nets, and wicker traps led to a few settled communities in parts of Southern Australia. Such communities maintained systems of hand-constructed dams, reservoirs, and channels in an enormous patchwork of wetlands running along rivers to the sea. These aquaculture systems ensured rich fishing grounds, even in times of drought. The larger populations supported by these innovations led to a thriving trade network. Eventually, incremental improvements might have generated complex societies with more advanced technologies. It's not impossible to imagine that, left alone, Aboriginal Australians might someday have set out to explore the world and search for the great "Terra Borealis Incognita" that some Australian geographers believed kept the celestial sphere in balance.

The other major expansion of humans into new territory was the settlement of the Americas. Although scientific consensus holds that the Americas were settled from Asia, the pattern of migration, its timing, and the exact place of origin of the settlers remains a topic of vigorous debate. As children read in their school textbooks, the leading theory is that migrating people walked into Alaska from Siberia around fourteen thousand years ago, on either land or sheets of ice, following herds of migrating animals they were hunting. This would have been possible during the last glacial maximum, an ice age that profoundly cooled Earth's climate, causing drought, greatly expanded deserts, and lowered sea levels. With seas as much as 400 feet (about 125 meters) shallower, land connections would have joined Britain to Europe and Siberia to Alaska, and bound most of the Indonesian islands together.

There's one problem with the theory of walking from Siberia to Alaska. During the last glacial maximum, enormous ice sheets several kilometers thick covered the northern continents.

The scale of such ice is utterly unfathomable to us today: the tallest skyscrapers wouldn't have come close to peeking above the top. Chicago's great ice sheet would have been three times thicker than the height of the Willis (formerly Sears) Tower. Toronto and Montreal boasted ice cover five times thicker than that. These continent-sized glaciers were heavy enough to carve out the major North American waterways, including the Great Lakes. Essentially all of Canada, Alaska, and the entire Siberian plateau would have been covered with massive ice sheets. Ice doesn't support plants or provide nourishment for migrating herds, let alone people. It hardly seems likely that humans walked over such glaciers into North America.

They probably didn't. Although Alaska today supports glaciers encroaching on the ocean, the coastline is relatively temperate and mild. During the glacial maximum, the entire coastline would have moved hundreds of miles southward into the Pacific. There may have been an ice-free corridor along the coast. Even if land access was blocked, there would have been ice-free patches and islands supporting vegetation. Moreover, the kelp forests along the Pacific coast are highly productive environments, providing a rich array of fish, mollusks, seabirds, seals, walruses, and otters that supports many human communities today. Thus, it's likely that our image of nomadic wanderers following herds of grazing animals might have to be revised; picture, rather, fishermen hunting and gathering their way along the shore, or perhaps using boats to hop from island to island down the coast.

Even if an ice or land bridge never existed, it's likely that the Americas would have been settled anyway. We often forget just how narrow the Bering Strait actually is. The former governor of Alaska may not be able to see Russia from her house (as she was quoted in parody), but on a clear day you really *can* see Russia from the US mainland in Alaska. It's less than fifty miles away, and counting islands, you'd never have to cross more than twenty-five miles of open water. Someday there might be a high-speed rail link (via bridge or tunnel) between the United States and Russia. In fact, another human migration from Siberia to North America definitively took place, much later, exclusively by sea. The ancestors of the Inuit arrived relatively recently in a separate wave of colonization from the original Native Americans, crossing by boat from Asia in the last few thousand years. They maintain contact with their Chukchi brethren in Siberia to this day.

Around a thousand years ago, the Inuit spread across the Canadian Arctic, displacing cultures who'd lived there since the first wave of American settlement. The Inuit brought with them far more advanced technology from Asia, including kayaks and umiaks,* large watertight boats of animal skin that were more seaworthy than wooden rafts or dugout canoes. They also brought more advanced tools and weapons than prior Arctic cultures and were able to fashion highly effective stone knives, spear tips, darts, and harpoons that kept a sharper edge. Their arsenal included copper and iron harpoons and lances used to hunt giant bowhead whales weighing up to one hundred tons.† Inuit metal, like that of other pre–Bronze Age civilizations, was collected from meteorites. (Space debris is more easily spotted in the Arctic because it tends to stand out against a white carpet of snow and ice.) The Inuit also had another huge advantage over earlier inhabitants of the region—dogs. Dogs are widely used by the Inuit to pull sleds, guard homes, and track prey. With their more advanced technology, within a few centuries, the Inuit had wiped out or displaced the original native inhabitants of the Canadian Arctic.

Sound familiar?

Vikings may have been the first Europeans to land in North America, but it's hard to argue that they were the first technologically advanced Eurasians to do so. The main distinctions may be that the Vikings used writing, had the ability to forge metal from ore, and kept domesticated animals for food. These distinctions seem insignificant in the context of small bands of Vikings, so I think we must surely award the Inuit the dubious honor of being the first Old World people to explore, settle, and conquer

---

* "Kayak" means "man boat" and "umiak" means "woman boat"—similar, but one for men, the other for women.

† Other important Inuit technologies: snow goggles made of bone (they look like futuristic sunglasses), hooks, needles, awls, shovels, ice picks, and sealskin buoys to recover stricken whales on the hunt.

16

in the Americas following its initial settlement. By the year 1300, Inuit arrived in western Greenland and started moving down the coast, entering eastern Greenland by 1400. This timing coincides perfectly with the demise of Viking settlements in those regions. It's a near certainty that competition with the Inuit, together with the challenge of dealing with a cooling climate, contributed to the Vikings' demise. The Inuit possessed far better mastery of the Arctic environment; for example, they built superior boats for navigating ice floes and were able to hunt large animals at sea. So these former Eurasians may also hold the distinction of being the first and only North American people to decisively eliminate a European rival.

We often picture Native Americans before European contact as small scattered bands of hunter-gatherers, but that's not remotely accurate. Fourteen thousand years is a long time to occupy continents: in less than a quarter of that time, Europe went from hunter-gatherer bands to the modern age. During their isolation, Native Americans diverged into dozens of language groups and hundreds of unique cultures. By the time of Europeans' arrival, most Native Americans lived in settled agricultural communities. There were large cities, too, not only in Mesoamerica and the Andes, but also throughout the continental United States. Estimates put the population of the Americas at well over a hundred million by the arrival of Columbus. In 1492, there were probably more people living in the Americas than in Europe.*

The main reason we think of Native Americans as nomadic hunter-gatherers is that up to 90 percent were killed by disease before seeing a European. For thousands of years, Native Americans had evolved in isolation from Europeans and Africans. It's not that Europeans were completely immune to disease. The Black Death killed up to a third of the population of Europe between 1347 and 1351, and this was only one of many waves of pestilence sweeping across Eurasia. However, this constant exposure meant

---

* Some parts of the Americas only recently returned to pre-Columbian population levels. For example, Mexico didn't reach its pre-conquest population of around forty million until the 1960s.

that surviving Europeans developed stronger immunities. Diseases tended to originate in Eurasia and Africa as a result of living in close proximity to livestock, the source for an estimated 75 percent of all human pathogens. The Americas had no large domesticated animals except llamas, who lived in smaller numbers with less human interaction, and no microbiological contact with the Old World until Europeans arrived.

Another reason Native Americans were so susceptible to disease was their genetic similarity. The more immunologically diverse a population, the harder it is for a disease to spread through it because the disease has to adapt to differences. Because North America was settled by a tiny initial population from Asia, that population's descendants possessed low genetic diversity. That's why, for example, whereas most blood types are distributed throughout the world, Native Americans are almost exclusively type O. Genetic similarity allowed European diseases to cut swaths through native communities.

In addition to European diseases, there were plenty of African ones, too. When Native Americans died in droves, European colonists turned to African labor in the form of slavery. Slavery was, of course, widespread throughout the American South, but it was even more pervasive in the Caribbean and South America, especially Brazil, where over four million Africans were brought to work on sugar plantations—almost half the total brought to the Americas. More Africans came to the Americas between 1500 and 1800 than Europeans, the vast majority in chains. These Africans brought with them diseases such as malaria and yellow fever, which found new homes in the American tropics. Thus, Native Americans faced a double onslaught from both European and African diseases.

However, the biological warfare wasn't entirely one-sided. There's at least one example of a disease that is thought to have traveled in the opposite direction. Syphilis was indisputably present in the Americas before European contact, but it was first recorded in the Old World in Naples during the winter of 1494 to 1495. The Italians claimed that it was introduced and spread by French troops; hence, it was often known as the "French disease."

Syphilis has long been a nuisance in Europe, but for the first hundred years, it was deadly. In an outbreak called the "Great Pox," syphilis killed as many as five million Europeans before mutation and natural immunity reduced its virulence.* The disease's sudden appearance in Europe immediately following the discovery of the New World seems unlikely to be a coincidence. Rather, syphilis was probably carried from the New World to the Old, possibly as early as Columbus's first voyage, either by one of the captive natives he returned with or by one of his crew—several of whom participated in the invasion of Italy where the disease was first reported.

After so many centuries apart, humanity was probably fated to suffer tragedy in the process of reunification, but it was Native Americans who bore the brunt of the catastrophe. European observers noted the immense numbers of indigenous people who began dying from initial contact, but the sheer scale of the devastation was overlooked because once diseases were introduced, epidemics raced far ahead of European explorers, depopulating entire continents. Disease disproportionately impacted Native Americans until the twentieth century, but the vast majority were probably wiped out within a hundred years of first contact, far from European eyes. Once diseases had depopulated the continents, new European arrivals moved into the vacuum and assumed that there'd always been relatively few indigenous people in the Americas.

---

* Since syphilis is spread exclusively through sexual contact, this also gives us insight into the private lives of Europeans of the time. One would think that in a more religious and sexually repressed society, STDs wouldn't be able to spread so rapidly. But that never seems to work out in practice.

3 | PEOPLE OF THE SEA

Prior to such inventions as writing, maps, and technologies that could transport people around the globe, geographic knowledge remained fragmented and compartmentalized. However, this shouldn't be taken to mean that earlier explorers were any less daring or skilled. Some were capable of astonishing feats, all the more impressive for their lack of written means of communication. Perhaps the most remarkable exploratory feat humans have undertaken is the settlement of the Polynesian triangle between Hawaii, Easter Island, and New Zealand—representing an area equivalent to the size of Africa, or almost four times the size of the United States—by a single culture, in little more than a thousand years. Relying on innovative ship designs, and taking along their animals, crops, people, and culture, the Polynesians managed to settle virtually every speck of land across the largest ocean on our planet.

The Pacific is vast. Covering a third of Earth's surface, it could comfortably fit within its borders all seven continents were they pressed together. When Europeans began to explore the Pacific, they were amazed to discover that almost every island was already inhabited. To European eyes, the local watercraft looked hopelessly unsuited for long sea voyages. In some places, such as Easter Island, the natives had no apparent seafaring skills and no more than flimsy canoes cobbled together from driftwood. Yet, somehow these "primitive" people had, not once, but many times,

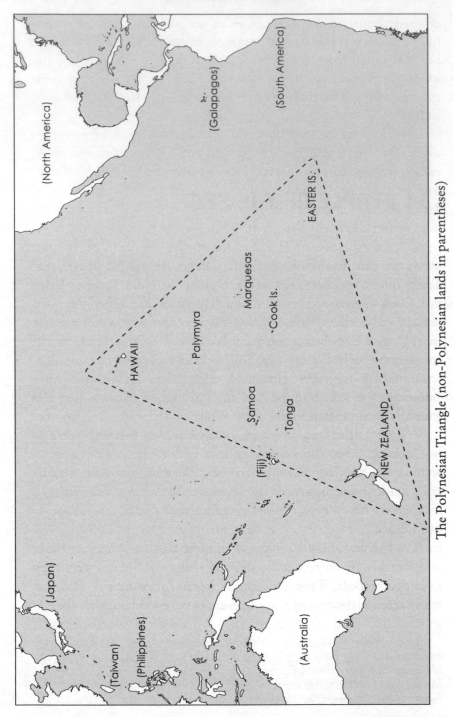

The Polynesian Triangle (non-Polynesian lands in parentheses)

crossed thousands of miles of open sea to land safely at veritable pinpricks in the largest ocean on Earth. It was as if Columbus had sailed from Spain thousands of miles across the Atlantic by canoe to make landfall on the tiny island of Bermuda, without even knowing it existed—and then repeated the trick hundreds of times.*

The mystery of how this was possible spawned bizarre theories. One held that the Pacific was once a giant continent. A cataclysm, perhaps a supervolcano or earthquake, must have submerged most of the land, leaving only a few islands above water. Thus, Polynesians were actually survivors of a terrestrial people who found themselves isolated in hundreds of communities scattered across the ocean. This catastrophe must have been a recent event, as Europeans observed that Polynesians still shared the same culture and language. When he reached New Zealand, Captain Cook summoned his Tahitian navigator and translator, Tupaia, who, with nothing to lose, attempted to communicate with the local Maori. To everyone's utter amazement, Tupaia and the Maori spoke essentially the same language, despite coming from islands separated by over three thousand miles. It was as if a ship from Barcelona had sailed up the Persian Gulf to discover that the local Iranian inhabitants were native speakers of Spanish.

Polynesian forebears probably originated on the island of Taiwan around three thousand years ago. Leaving before developing writing but after adopting agriculture, they brought along breadfruit, taro, yams, bananas, and coconuts to plant at islands they discovered. They kept domesticated pigs, chickens, and dogs for food, but the majority of their protein came from the sea in the form of fish and shellfish. They were experts at long-distance trade and found eager customers for flint and obsidian, which were used in making tools. Though lacking writing, Polynesian cultures maintained elaborate oral traditions. Hawaiian chiefs, for exam-

* Europeans had such difficulty finding Bermuda (isolated far out in the Atlantic) that most early expeditions sailed right past it without noticing. The island was only discovered by accident when ships blundered into it, as happened to a nine-ship 1609 English expedition to Jamestown that washed up on the reef, inspiring Shakespeare's *The Tempest*.

ple, learned to recite a list of their ancestors from memory for over twenty generations, tracing their lineage further back than most European royal families.

By around 900 BCE, Polynesian ancestors reached Tonga and Samoa, where their distinctive seafaring culture emerged. The exact dates are uncertain, but waves of Polynesians then spread north to Hawaii by around 300 CE, east to Easter Island by around 700 CE, and south to New Zealand by around 1300 CE. Although Polynesia contains thousands of islands, and some archipelagoes contain hundreds (Fiji has at least 332), most are extremely remote and well out of sight of each other on the clearest of days. Easter Island, for example, is one of the loneliest places on Earth, 1,200 miles from the nearest inhabited island (roughly the distance from Africa to Brazil). Polynesian navigators couldn't simply have picked a random direction in which to sail in hopes of finding land. Magellan, who in 1521 became the first European to sail across the Pacific, passed through the entire breadth of Polynesia from east to west without spotting a single island. So how could Polynesians manage to find essentially every habitable island in this largest ocean on Earth?

Another European theory was that Polynesian sailors discovered new lands accidentally, by getting blown off course. While this may have happened from time to time, "exploration by getting lost" couldn't possibly account for the bulk of Polynesian expansion. You'd prepare very differently for a sea voyage if you were embarking on a day trip as compared with a thousand-mile transoceanic journey.* Any Polynesian sailor who managed to get so drastically lost would soon be in serious trouble as they ran out of food and water. Even if some islands were located in this way, this methodology implies that accidental explorers simply retraced their steps and returned home to spread news of their discoveries—except, if they *were* able to do that, were they really lost? Instead, we must conclude that Polynesian settlers deliberately

---

* Interesting to reflect that "journey" connotes significant travel even though the word comes from the distance you can travel in a day, from the French *"journée."*

traveled to new lands in large oceangoing vessels with their families, livestock, and seedlings.

This conclusion is compatible with Polynesian naval technology, which, despite its basic appearance, is extremely well suited for long sea voyages. Polynesian outrigger canoes can be very large—up to fifty feet long—and support a crew of twenty or more for weeks out at sea. Picture a long canoe with perpendicular cross-arms holding mini canoelike "outrigger" floats ten to fifteen feet out, parallel with the main canoe. Smaller canoes use a single outrigger on one side, whereas larger canoes often mount outrigger floats on both sides for greater stability. Some versions have two canoes lashed to the cross-arms forming a catamaran. Unlike normal canoes, an outrigger is very stable on the ocean, even in heavy waves, blending the stability of a wide boat with the agility of a narrow one. They're still popular for racing in Hawaii, Tahiti, Samoa, and New Zealand.

Each Polynesian island cultivated a class of esteemed navigators who not only discovered remote islands thousands of miles apart but also maintained a continuous trade network between them over the centuries (although some, like Easter, eventually lost contact). Navigators sailed between islands using knowledge passed down through oral tradition from master to apprentice, often in the form of song. To find direction, they memorized the motion of stars, the angles of the Sun and Moon, the patterns of clouds clustered over islands, and the ebb and flow of ocean currents. Navigators would look for changes in wave patterns that signaled disruption by land to help them determine the size and direction of island chains. They probably also followed bird migration* and sometimes brought along their own caged frigate birds to extend their view over the horizon. When released, the bird would rise up and fly toward any land in sight, or return to the canoe to avoid getting wet.

---

* This idea was suggested by the Australian navigator Harold Gatty, who wrote a survival guide for the US military during the Second World War that outlined Polynesian navigation techniques to help Allied servicemen stranded at sea. He also was navigator on one of the first airplanes to circumnavigate the world (1931), relying almost entirely on dead reckoning by compass.

On his first Pacific voyage, Captain Cook recruited the Polynesian navigator Tupaia. Even though Tupaia had never seen a map, he was able to draw from memory a chart of all the major islands within two thousand miles to the north and west of his home island of Raiatea (near Tahiti). This is akin to reciting directions from Detroit to Los Angeles in a time before roads. Tupaia recalled the existence of 130 islands and named 74 on his chart. Revealingly, even though Tupaia could describe these islands, he'd visited only a few. His island's navigators had not been to the most distant since his grandfather's day, but their knowledge had been passed down through the generations. Around the campfire, they'd sing about lands to the west, embellished with tall tales to enhance the recollection.

Polynesian navigators didn't recklessly sail out to sea. They'd generally try to sail against the wind, so that they could always return safely if they encountered trouble. On long voyages, navigators were segregated so they could couldn't influence each other.* Sitting apart from the rest of the crew (as much as possible on a canoe at sea), navigators continuously concentrated on how far they'd gone, where they were, and where they were going. They were usually forbidden to sleep (on short voyages), lest they lose their reckoning. During the day, they'd track direction based on the Sun. Even on an overcast day, they could estimate the Sun's position from prevailing light seeping through the clouds. At night, they'd take their bearings from the Moon and stars.

Polynesians weren't only navigators but pioneers, sailing in large multicanoe colonial expeditions with families, dogs, chickens, and pigs. They collected rainwater along the way, and also supplemented their food and water with fish (ocean fish are a great source of drinking water because their kidneys separate out the salt). They brought seedlings of breadfruit, taro, yams, and bananas; a large assortment of tools; and high-quality stone to fashion whatever additional tools they might need.† This was no

---

* This concept of comparing multiple independent estimates is the key to error correction in modern guidance systems that steer ships, airplanes, and even rockets to space.

† The prehistoric version of 3-D printing.

mere act of getting lost—colonizing Polynesians were far better equipped to settle a new land than many of their later European counterparts who arrived in the Americas with only two options: starve or plunder.

Over time, Polynesians diverged into a diverse array of inter-related cultures spanning thousands of islands, from Samoa and Tonga to the Cook Islands and Marquesas. In Fiji, Polynesians blended with earlier Melanesian cultures. To the north, they became Hawaiians. To the south, they became the Maori of New Zealand. Remains of settlements extend to the Auckland Islands, three hundred miles south of the southern tip of New Zealand, where temperatures regularly drop below freezing in winter (June–September). Traces of artifacts, possibly Polynesian in origin, have been found on remote Macquarie Island, lying a further three hundred miles south, halfway between New Zealand and Antarctica. If authentic, these would seem to support the Maori legend of a great navigator who led an expedition south until they reached "a place of bitter cold where structures rose from a solid sea." This description appears to match the Ross Ice Shelf, or possibly the Antarctic mainland, but may equally represent floating icebergs. Either way, Polynesians ventured farther south than anyone on Earth before the Age of Exploration.

In the years between 700 and 1100 CE, Polynesians reached the far eastern limit of their expansion at the remote outpost of Easter Island. There, they found a fertile island filled with towering forests, including a giant now-extinct relative of the Chilean wine palm. This largest known palm would have stood a hundred feet tall, with five-foot-thick trunks. Easter's settlers probably arrived from the Marquesas Islands, over twelve hundred miles to the west. They initially prospered, establishing a civilization of some twenty clans and villages, with a population reaching around fifteen thousand. Given the total island area of sixty-three square miles, this implies a population density of 238 people per square mile—greater than Hawaii's today, which couldn't survive without food imports from the wider world. Such a concentration would surely have caused ecological problems.

The most visible remains on Easter Island are 887 *moai*, "giant head" statues scattered around the island. Some are of a staggering proportion. The tallest, "Paro," stood over thirty feet high with an exaggerated head about the same size as the Statue of Liberty's. The heaviest weighed 86 tons, as much as twenty full-grown African elephants. One unfinished statue was of even grander proportions, with a projected height of up to sixty feet and weight of up to 270 tons. Just producing, let alone transporting, these incredible statues must have required a complex society. Yet, magnificent though they were, all the statues had been toppled by the time Europeans discovered the island in 1722. It seemed the islanders had destroyed their own statues in response to some calamity.

As Jared Diamond explains in his excellent book *Collapse*, Easter Island is a classic example of overuse of a fragile ecosystem. Despite being productive, the soil was thin and susceptible to degradation once the trees were cut down. To compound the destruction, rats that accompanied the human settlers would have eaten the palm nuts, preventing regrowth. With the elimination of forests, the islanders could no longer build canoes, which not only curtailed their ability to fish but also irrevocably cut off external contact. The struggle for survival fanned the flames of constant warfare and persistent cannibalism. By 1722, Europeans found fewer than 3,000 islanders eking out a meager existence. Diseases and slavers killed or abducted most, so that by 1877 only 111 survivors remained. Although the population has made a significant recovery, it's entirely dependent for its survival today on government support and tourism. We should take this as a warning of what can happen when a confined but growing population outstrips its resources, lest the fate of Easter Island play out on a global scale.

Polynesians weren't the only early seafarers to cross oceans. While Polynesians were migrating across the Pacific, their cousins from the Sunda Islands of Indonesia embarked on impressive voyages covering more than 3,500 miles in the opposite direction. Using similar navigational techniques, they explored the entire circumference of the Indian Ocean from Malaysia to South Africa, and established settlements in the Seychelles and Chagos Islands.* Especially active between 200 BCE and 500 CE (and thus, contemporaries of the Romans in the West), these "Polynesians of the Indian Ocean" operated a trade network stretching from Africa to Indonesia. Long before anyone had heard of the Portuguese, these ancient mariners carried spices like cinnamon and pepper bound for Europe.

The legacy of these seafarers survives on the island of Madagascar. Although we evolved in Africa and lived exclusively there for 97 percent of our existence, early humans never made it to Madagascar, which has been separated from the mainland for 160 million years. Since Madagascar was isolated for so long, the island was full of unique and fascinating creatures, and still is. Over 75 percent of all animals on Madagascar exist nowhere else on Earth. In the absence of monkeys, lemurs exploded to fill a dizzying array of ecological niches. In turn, fossa, which are reminiscent of small mountain lions but are actually related to the mongoose, evolved to hunt lemurs. A hundred species of chameleon call the island home, more than half of all types on Earth. When humans first arrived, they were greeted by elephant birds standing ten feet tall and weighing a thousand pounds. Laying eggs three feet wide (equivalent to 6 ostrich or 160 chicken eggs), elephant birds could, in a single act of reproduction, furnish an omelet capable of feeding an entire village—and often did, judging from shell fragments found at archaeological sites. Pursued by hungry hunters and egg gatherers, these gargantuan birds were driven to extinction four centuries ago.

Today, Madagascar is a melting pot of cultures from Africa, Arabia, India, China, and Europe, but its first settlers were seafarers who arrived over a thousand years ago in outrigger canoes. The island still shares many cultural connections with Polynesia,

---

* The Chagos Islands sit near the equator right in the middle of the Indian Ocean. British since the Napoleonic Wars, the largest island of Diego Garcia has been leased by the US military since 1966. Due to its strategic location, it has become the leading satellite tracking station and military base in the region, projecting B-52 strikes as far as Afghanistan.

including linguistic roots, music and dance, "burial" of the dead in canoes at sea, and the cultivation of taro, bananas, and coconuts. These are just a few of the remaining legacies from voyages of discovery spanning from Africa through Southeast Asia to the Pacific. Long before the arrival of Europeans, skilled navigators were exploring and settling a vast oceanic domain spanning a quarter of our planet.

4 | ANTIQUITY

The Mediterranean, literally a "sea at the middle of Earth," as its name implies, was once a vast salt plain. During the Messinian Salinity Crisis six million years ago, Eurasia collided with Africa, sealing the outlet to the Atlantic Ocean. As seawater evaporated, salt dunes piled up in thick layers, and most marine life perished. Then around five million years ago, a sudden breach opened to form the Strait of Gibraltar between Spain and Morocco, and the Atlantic rushed in. For a time, this would have been the tallest and most impressive waterfall in the world, with a thousand times the outflow of the Amazon falling almost a mile to fill the dry seabed.* Shortly thereafter, the first human ancestors would have reached the shores of the newly formed sea—one separating Europe, Africa, and Asia and destined to become a superhighway of cultural exchange.

Humans have a long history with the Mediterranean. Stone tools found on the island of Crete date back 130,000 years, indicating that early humans (and likely Neanderthals, too) used primitive boats to sail its waters. Centrally located, with rivers pouring in from all directions, the sea became a commercial hub. It was

---

*A similar flood probably formed the Black Sea. Isolated from the Mediterranean for tens of thousands of years until around 5,600 BCE, rushing water filled the Black Sea in a sudden breach of the Bosporus. This event, in a region thickly populated with pre–Bronze Age villages, may form the basis for flood myths like Noah's.

calm compared with the open ocean, so sailors could travel great distances in relative safety hugging the shore, always in sight of land. The Mediterranean's east-west alignment meant a uniform climate, encouraging the spread of plants, animals, people, and civilizations. Its dry, temperate weather was ideal for growing grains, grapes, and olives, and herding goats and sheep. In ancient times, from Spain to Italy to Greece to Turkey to Egypt to Morocco, "Mediterranean" meant "the region where olives grow."

At the southeastern corner of the sea, Egypt built its cradle of civilization around the fertility of the Nile. The Nile is life in Egypt, bringing not only fresh water but annual flooding that replenishes the soil with mineral silt from the highlands of Ethiopia. Without the Nile, Egypt would be an uninhabitable desert. Even today, Egypt is utterly dependent on the river. Supporting the third-largest population in Africa (over ninety million), the Nile is within walking distance of more than 95 percent of Egyptians. Sustained by the Nile's bounty, Egypt developed agriculture, writing, pottery, masonry, and law, and founded a series of dynasties that ruled the southeast corner of the Mediterranean for more than three thousand years. Egyptian architects built such monumental structures as the pyramids, the only remaining wonders of the ancient world, and the Sphinx, still the largest statue ever carved from a single rock.

The ancient Egyptians are not generally renowned as exceptional explorers. Partly, this is the result of their mythology, which held that reincarnation required proximity to the life-giving Nile. Thus, dying in a far-off land carried not only mortal but immortal danger. Nevertheless, the strategic location of Egypt made it an important commercial hub. Like North America, Egypt sits astride two major global seas that lack a natural water connection. Long before the Suez Canal opened in 1869, the Canal of the Pharaohs connected the Red Sea to the Nile, and thence to the Mediterranean. This meant that a boat could theoretically sail all the way from the Mediterranean to the Far East through Egypt, without having to go all the way around Africa. The canal was difficult to use, facing the constant problem that salt from the Red Sea would invade the Nile, killing crops and ruining fresh water. However,

Greek engineers in the time of Ptolemy I Soter (an ancestor of Cleopatra) solved this problem with our modern solution: canal locks that carried boats up and down across changes in elevation.*

There is plenty of evidence that ancient Egypt organized long-distance sea voyages, but we don't know exactly where. Ancient Egyptians lived a very long time ago. The pyramids were older to Cleopatra than Cleopatra is to us, built over 4,500 years ago at a time when woolly mammoths still walked on Earth. Since Egypt predates most other civilizations, it's unclear exactly who they would have visited. Greece was filled with pre–Bronze Age nomads and goat herders. Italy, France, and Spain were even more backward. The Middle East supported some civilizations, and there is evidence that Egypt traded with the eastern Mediterranean to acquire cedarwood (featured on Lebanon's flag today), but most of the Mediterranean was a backwater wilderness. Egyptians may well have visited much of it, but no records of their westward travels have survived.

We do know that they spent significant energy traveling in the other direction, down the Red Sea to a land they called Punt. We don't know for sure where Punt was, but based on descriptions it was probably somewhere near the Horn of Africa, where the Red Sea meets the Indian Ocean. This is also supported by genetic studies of mummified baboons brought from Punt to ancient Egypt, which eventually ended up in the British Museum.† The distance from Egypt to the Horn of Africa is around two thousand miles, which is quite impressive for ancient travelers. The earliest recorded expedition to Punt was organized during the Old Kingdom, around the time the Great Pyramid was constructed (2560 BCE). By the Middle Kingdom (c. 2050–1650 BCE), expeditions to Punt were common. Punt doesn't seem to have been an equal to Egypt in terms of power or cultural achievement, but it was none-

* The Nile–Red Sea canal eventually silted up and was abandoned.

† It was common for Egyptians to mummify animals, especially cats, dogs, monkeys, and birds. This apparently had nothing to do with religious ritual, but, rather, was done in memory of a beloved pet. If you want to remember Fluffy, why not just keep him on the mantel, where you can continue to enjoy his company?

theless a wealthy kingdom that maintained trade connections into the heart of Africa.

The cultural impact of these overseas voyages is preserved in one of the earliest known works of fiction from anywhere in the world. Over four thousand years old, and recorded on a papyrus scroll, *The Tale of the Shipwrecked Sailor* tells the story of an Egyptian castaway on a deserted island. There, he meets a serpent who turns out to be the "Lord of Punt," possessing magical powers. The sailor pledges to tell the pharaoh about the power of the serpent and spread news of his existence far and wide. In exchange, the serpent bestows on the sailor precious gifts of incense, spices, elephant tusks, live baboons, and greyhounds. Upon his return, the pharaoh is so pleased with the gifts that he raises the sailor to nobility and showers him with riches. This story bears all the hallmarks of an explorer's adventure fable: an inexperienced hero finds an enchanted land, confronts and overcomes a monster, and survives wiser for the experience. Clearly, not just daring voyages but *tales* of these voyages were as much in the popular consciousness in ancient times as they are today.

By the reign of Queen Hatshepsut* around 1500 BCE, Egyptian ships regularly crossed the Red Sea. They traded for copper, bronze, precious stones, and other goods transported overland down to Eilat at the head of the Gulf of Aqaba.† Hatshepsut herself accompanied the most celebrated expedition to Punt, described in hieroglyphs at the temple at Deir el-Bahri. The fleet carried not only gold to purchase goods but also a strong contingent of soldiers intended to demonstrate the power of Egypt. By this time, the Egyptian realm had expanded down the Nile to mix with the Nubian civilization.‡ This brought extended contacts into the

---

* One of the earliest female rulers in history, after the short tenure of Pharaoh Sobekneferu a few centuries earlier and a Sumerian queen named Kubaba.

† Still a strategic port today as Israel's only access to the Red Sea, and also the site of Lawrence of Arabia's famous attack on the Turkish garrison in both the Hollywood and historical versions of the story.

‡ The Nubians were distinct but similar to the Egyptians. They actually built more pyramids, so there are more of these in Sudan than Egypt (though the Sudanese pyramids tend to be smaller).

African interior. We're not sure how far Egyptians traveled, but they at least recorded the split of the Nile into two tributaries at Khartoum, a thousand miles south. Later expeditions during the Ptolemaic period traced the Blue Nile as far as the Ethiopian Highlands, beating European explorers who "discovered the source of the Nile" by over two thousand years.

There are other clues of Egyptian forays deep into Africa. As depicted in murals, Egyptian princesses used a bluish-white metallic substance as eye shadow, and chemical samples from mummy caskets have confirmed this to be antimony. The only definitively known source of antimony at the time was a set of open mines in southern Africa. Another clue is the case of the okapi. Okapis are relatives of giraffes that look to be crossed with zebras thanks to stripes on their legs. They're rare today but would have been more common back then. Nevertheless, okapis are not suited for the open plains of Sudan and wouldn't have roamed far beyond the jungles of central Africa, more than two thousand miles from Egypt. Despite this, okapis were known to ancient Egyptians and even figured in their mythology. These examples aren't conclusive, and both antimony and okapis may have been obtained indirectly through trade. Yet, even if this is the case, it suggests that ancient Egyptians had connections extending far deeper into Africa than Europeans until the Victorian age.

Around 1200 BCE, a vigorous new seafaring civilization rose in the eastern Mediterranean. This wasn't a terrestrial empire ruled by pharaohs but a confederation of city-states linked by cultural ties. Over time, these "Phoenicians" developed the world's first thalassocracy—a maritime realm based on a network of seaborne trade routes.* Each Phoenician city was politically independent, and it's uncertain to what extent they considered themselves a single people. In this way, they were like the Greek city-states, sharing a common culture and often aligning in times of crisis but also

* "Phoenician" comes from the Greek for "purple," based on a dye derived from crushed seashells that the Phoenicians famously traded. This highly prized ancient dye is why purple has been the royal color throughout history. We don't know what the Phoenicians called themselves, but they came from the land of "Kenaani" ("Canaan" in the Bible).

fiercely protective of their individuality. The Phoenicians' language introduced one of the first letter alphabets—the ancestor of the Greek and Latin alphabets used by all European languages today. Indeed, many of the cultural precepts of Western civilization originated with the Phoenicians and spread across their maritime trade routes.

Phoenicians were first to colonize the western Mediterranean in a major way. Without their influence, Western civilization probably wouldn't exist, or at least would be very different. They built trading posts and towns, establishing networks that were eventually inherited by Rome. They sold glass, textiles, and their famous purple dye to the early Greeks, and wine and timber to Egypt. As the Greeks adopted the maritime traditions of the Phoenicians, the two cultures effectively split the Mediterranean, with Phoenicians dominating the south shore from North Africa to Spain, and the Greeks colonizing Italy and southern France. In the center, Sicily split into a Phoenician west and a Greek east. This balance of power lasted until the rise of Rome, at which point Rome conquered Greece and faced off against the Phoenician-derived civilization of Carthage for control of the sea.

Pioneering the development of large oared and sailing ships, Phoenicians laid the groundwork for all subsequent European naval technology. Explorers for the first time ventured out of calm Mediterranean waters into the Atlantic beyond the Strait of Gibraltar. By 500 BCE, they were sailing in sturdy oceangoing vessels to the island of Britain to barter for tin and lead. When Rome was still a tiny village struggling for survival, the Phoenicians were establishing trading posts and mines in Cornwall. Around 425 BCE, the Phoenician colony of Carthage dispatched an explorer named Hanno the Navigator at the head of a fleet of sixty ships to explore the Atlantic. He landed in the Canary Islands and possibly Madeira, a thousand miles into the Atlantic. After founding several trading posts in Morocco, he sailed down the coast of Africa. It's unclear exactly how far Hanno ventured, but based on descriptions of the voyage, he is reckoned to have reached at least as far as modern Senegal, where Africa extends to its westernmost point, and may have sailed as far as the jungles of Central Africa.

Either way, Hanno traveled farther than any Europeans until the Portuguese almost two thousand years later. Recounted by the Greek historian Herodotus, and depicted on murals in the ruins of Carthage, Hanno's voyage featured encounters with natives, volcanoes, and exotic animals that may have been gorillas. Herodotus describes the "silent trade" that the Phoenicians conducted as they sailed down the coast of Africa. To surmount the language and trust barrier, Phoenician crews would find a beach near local villages and lay out goods for trade in rows along the sand. The natives apparently realized that these goods were meant for trade and would lay out other goods next to them at what they considered a fair exchange. Then the crew would add or remove wares, negotiating, until a barter price was mutually agreed.

As with the Egyptians, we don't know the full extent of Phoenician travels because most records from the ancient world were lost during the intervening millennia. Famed in antiquity for shipbuilding skills, the Phoenicians built better seagoing vessels than any in Europe until around the time of Columbus. They invented big-bellied multitiered ships with keels, rope-and-tackle-hoisted sails, and watertight caulking between planks. Their largest ships may have displaced as much as 450 tons—more than Columbus's largest ship, *Santa María*. These were a people who were used to long sea voyages. Some have gone so far as to suggest that Phoenician navigators crossed the Atlantic to the Americas. While this would have certainly been within their capability, there's scant evidence to support this claim, so we must assume that they didn't, or at least not intentionally. Nevertheless, given that they had the ability to do so for a duration of more than five hundred years, we can't completely rule out the idea of a crossing to the Americas, even if by accident.*

There is at least one surviving record of an early long-distance sea voyage undertaken jointly by Egyptians and Phoenicians. As recounted by Herodotus, the pharaoh Necho II is supposed to

---

* Artifacts of supposed Phoenician origin have popped up in the Americas over the years, including the 1889 "Bat Creek" tablet found in a burial mound in Tennessee; however, so far these have all proved to be likely forgeries or misidentifications.

have hired a Phoenician crew for an expedition to circumnavigate Africa around 600 BCE. In *The Histories*, Herodotus recounts that the fleet sailed down the Red Sea past the Horn of Africa and around the continent to reenter the Mediterranean at Gibraltar. As the crew rounded the southern tip of Africa, something strange happened. As they took a westerly course, they observed the Sun passing through the sky from sunrise (behind) to sunset (in front) on the right side of their ship (something that never happens in Europe). Most contemporaries took this as reason to doubt the account, and even Herodotus dismisses the detail as fabrication. In hindsight, however, it may actually confirm the story's veracity. Any crew sailing westward around Africa would indeed witness the Sun passing through the sky to their right because they would be south of the equator. The story also precisely captures the geographic fact that Africa can be sailed around. It seems unlikely that Herodotus would guess this by accident.

Is it really so hard to believe that ancient sailors could have accomplished something that was routine for Europeans during the Age of Exploration, especially considering that they would have been using essentially the same technology? All it would have taken is a brave crew of ancient adventurers and the will to explore the unknown. We may never know. The sample of surviving artifacts is too small to draw definitive conclusions. Many of the exploits of the ancient world were probably never recorded, while others have no doubt been lost over thousands of years. If a future civilization were judging our accomplishments by, say, deciphering the characters on the Lincoln and Jefferson Memorials, they might never imagine that we'd traveled to the Moon. We can definitively recount some exploits of ancient people and guess about others, but we can never know with certainty the full magnitude of their achievements.

The situation is worse than you might imagine. Ancient knowledge was treasured by the Greeks and Romans, but as Rome collapsed, much of it was lost. In 529, Emperor Justinian closed down the Greek schools of philosophy in a quest to unite the empire of the Eastern Romans under a single religion. The Library of Alexandria had housed over a hundred thousand scrolls, but this trove was destroyed in a series of lootings and fires, starting with Julius Caesar's siege in 48 BCE and culminating with a demolition decree by the Coptic Christian Pope Theophilus in 391 CE. We may never know the full scale of the destruction. Euripides is known to have written over ninety plays, but only a fifth of these survive. Of Herodotus's vast historical works, we have mere fragments recounted by others. Democritus first postulated the existence of atoms and is considered the founder of science, but all we have are intriguing secondhand whispers of his theories. There may have been other renowned scholars of whose very existence we'll never know.*

What we do know is that the people of ancient civilizations were enthusiastic explorers and scientists, as curious about their world as we are. One of the first curators of the Library of Alexandria was Eratosthenes (276–195 BCE), the preeminent scholar of his time. A true polymath, he seemingly dabbled in everything, from science, mathematics, and astronomy to poetry, music, political philosophy, and history. He even wrote plays. Eratosthenes's expertise was so diverse that contemporaries jokingly called him "Beta" (the second letter of the Greek alphabet) because he was second-best at everything. But in several fields it's clear that Eratosthenes was surely not Beta but Alpha. The father of geography, his masterwork *Geographika* mapped the known world, describing over four hundred cities and dividing Earth into five climate zones that we still reference: two freezing zones around the poles, two temperate zones, and the equatorial tropics.†

* Some works only survive because parchment was so expensive. Over the years, people would scratch out ancient writings to reuse the page, and imperfect erasure allowed the original writing to be reconstructed. Famous works of Archimedes, Cicero, and Seneca were all preserved in this way. It's a bit like rubbing out Hemingway to compose a grocery list.

† Some of his achievements include developing an algorithm for finding prime numbers (the Sieve of Eratosthenes), computing ballistics for catapults, accurately calculating the distance to the Sun, and not-so-accurately calculating its diameter (although a factor-of-four error isn't bad considering the Sun's size). Eratosthenes is who we can thank for fixing our calendar at 365 days, with a leap day every fourth year. He also invented a dating system similar to our own but starting with the conquest of Troy as year zero.

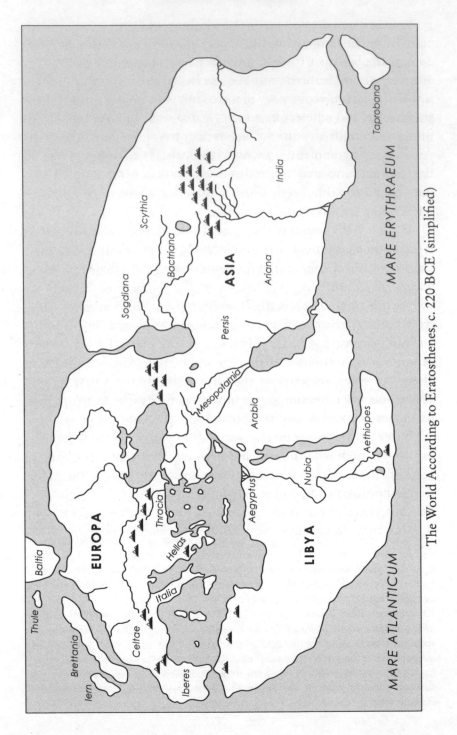

The World According to Eratosthenes, c. 220 BCE (simplified)

Eratosthenes created the first world map that we would recognize today. Encompassing Europe, Asia, and Africa, it accurately depicts the lands surrounding the Mediterranean, becoming progressively more featureless to the north, south, and east. The Nile is shown in its full course, even including the two lake sources that eighteenth- and nineteenth-century European explorers struggled for over a hundred years to find. The main rivers of Europe and Asia feature prominently, and in a time before the Roman Empire, Eratosthenes was able to sketch a rough outline of not only Britain but Ireland too. He overlaid the map with horizontal and vertical reference lines, inventing the concepts of latitude and longitude. More than two thousand years ago, Eratosthenes recognized that Africa was a continent that you could sail around, and that since the world was a spherical globe, you could get to Asia by sailing west.

Not only did Eratosthenes chart the world, but he also calculated its circumference, especially remarkable considering that he did so without leaving Egypt. A merchant's account had mentioned that at noon on the summer solstice in Syene (a town directly south along the Nile), an observer's shadow would block the reflection of the Sun when looking down a deep well, implying that the Sun was directly overhead on that date. For comparison, he measured the Sun's angle at noon on the same day in Alexandria. This turned out to be about one-fiftieth of a circle, or seven degrees. Assuming Earth is a sphere, he concluded that its circumference is fifty times the distance to Syene.* By this method, he computed the circumference of our planet with an error of less than 10 percent. Almost two thousand years later, Christopher Columbus knew

---

* We're not sure how Eratosthenes measured the distance to Syene. He may have simply asked local caravans how long it took them to get there. Carl Sagan suggests that he paid a man to walk there and pace out the distance. Either way, he came up with a distance of around 5,000 stadia. A stadium is about one-ninth of a mile, based on the length of the running track at the original Olympic stadium in Greece (which you can still pace today). Using stadia of 582 feet, this gives an Earth circumference estimate of 27,550 miles, within 10 percent of the accepted value of 24,900 miles. Repeating this calculation with a more accurate distance yields a result of 25,046 miles, an astonishingly small error of less than 0.16 percent.

about Eratosthenes's measurement but chose to believe a fanciful map that showed a circumference smaller by a third. Had it not been for this wishful thinking, Columbus would likely never have set sail—or if he had, he'd have correctly guessed that he'd arrived not in Asia but a new world.

5 | THE CLASSICAL WORLD

Around the time of ancient Egypt, a small but prosperous civilization was rising across the Mediterranean on the island of Crete. The Minoan realm was an early Bronze Age culture of merchants, artists, architects, and philosophers—the first link in a chain that would lead to Greece and Rome. They were seafarers, trading with Egypt as long ago as the construction of the pyramids, and colonizing the Syrian coast and Greek islands of the Aegean, sparking both the Phoenician and Greek civilizations. Then, suddenly, around 1400 BCE, the Minoans disappeared. Their decline was probably initiated by an earthquake or volcano that devastated Crete, destroying the palace of Knossos. The island was shortly thereafter overrun by mainland Greeks. However, Minoan culture wasn't so much destroyed by the Greeks as adopted by them. Soon the Greeks would make Minoan culture, religion, and art their own, and continue the Minoan maritime tradition.

Some of the best-known elements of Greek mythology are based on Minoans. In the tale of the Minotaur, King Minos conquers Athens and demands that every nine years, seven boys and seven girls be sent to Crete as sacrifices to the Minotaur, a monster with a human body but a bull's head.* The Athenian hero Theseus volunteers to take the place of one of the victims. Unrave-

* The original *Hunger Games*, proving that we never write anything new but just recycle old plots.

43

ling a string woven by Minos's lovestruck daughter, Ariadne, to mark his escape route, Theseus enters the Minotaur's labyrinth and slays the beast.* The labyrinth's architect was Daedalus, who was locked up in a tower to protect its secrets. A gifted inventor, Daedalus escapes with his son Icarus using wings of wax and feathers—an early tale of humanity's aspiration to fly. But Icarus flies too close to the Sun, his wings melt, and he plunges to the sea.†

The Greeks inherited the seafaring spirit of the Minoans. In the tale of Jason and the Argonauts, a band of heroes sails the Black Sea looking for a golden fleece from a winged ram (which became the constellation Aries). Another Greek classic, Homer's *Odyssey*, is perhaps the most widely appreciated explorer-adventurer story of all time. In penitence for sacking Troy (i.e., thinking up the wooden horse to sneak into the city), Odysseus and his crew endure a grueling decade-long voyage to lands unknown. First, the Lotus-Eaters give them fruits that sap their energy and make them forget their homes. Then they're captured by a Cyclops who plans to eat them. Next, a witch turns them into pigs, but she agrees to free them in exchange for Odysseus's love. Finally, they sail past the Sirens, who sing an enchanting song that lures sailors to their deaths. To avoid this fate, the crew plugs their ears with wax, except Odysseus, who is tied to the mast so he can enjoy their

---

* The Minotaur turned out to be Minos's illegitimate son, so Ariadne helped her boyfriend kill her brother. This doesn't seem nice, but everyone lived happily ever after. That is, except Theseus's father, who flung himself dramatically from some cliffs because the ship carrying Theseus home forgot to display a white signal as planned. And for some reason, Theseus was ordered by the goddess Athena to abandon Ariadne in Crete, so actually, no one was happy. Except Athena's brother Dionysus, who scooped up Ariadne in her misery. You really have to wonder if that was a setup on the part of the gods.

† After this, Minos pursued Daedalus to Sicily but was lured into a lethal trap of boiling water. Greek stories always seem to have gratuitous sex and violence, which may be the reason they've been such crowd-pleasers over the years. Think of them as *Game of Thrones* before HBO. In the Greek telling, Daedalus is portrayed as a mechanical genius, improbably credited with the invention of carpentry, which seems like something too broad and yet too basic to be invented by a single man. Later in life, Daedalus teaches his nephew Perdix his mechanical arts but, unable to bear a rival, allows him to plummet to his death one day while visiting the acropolis in Athens. Worst family vacation ever.

heavenly voices. Eventually, Odysseus and a few sailors stumble back to Greece, wiser for their voyage. Thus, "odyssey" entered our vernacular.

Perhaps more than any other culture, the Greeks invented Western civilization. Their most widely appreciated contribution is democracy, which began almost by accident when the citizens of Athens overthrew their tyrant in 508 BCE,* but their influence extends at least as much into the social spheres: drama, poetry, philosophy, literature, and the use of language.† It's remarkable to reflect that Greek masterpieces written two thousand years ago still brim with social relevance today. Sophocles, Euripides, Aeschylus, Aesop, and Homer cut to the soul of humanity with themes of power, humility, perseverance, integrity, and self-restraint.

Greek ideas spread through the Mediterranean over their maritime trading empire. Starting around 1000 BCE, the Greeks expanded beyond their Aegean home to found colonies across the Mediterranean and Black Sea. Coastal trading posts were set up to exchange goods with locals in the hinterland, who were themselves influenced by Greek culture and language. Eventually, many of these trading posts grew into the Mediterranean cities we know today. While Greek influence spread to France, Italy, and the Adriatic coast, the Black Sea became a Greek lake, dotted with colonies around its perimeter that exploited rich fishing grounds and salt beds. Within a few centuries, there were more Greeks living outside Greece than within it, in over five hundred overseas colonies. Many of these settlements outgrew their ancestral homelands, like Syracuse, Sicily, which became the largest Greek city in the world, surpassing even the population of Athens.

The Greek city-states and colonies were fiercely independent.

---

* Prior to this, Athens had harsh laws enacted by the ruler Draco (hence "draconian"). Voting in Athens was accomplished by a show of hands, or sometimes colored stones. Occasionally, citizens voted to banish troublesome people by scratching the person's name onto a pottery shard called an *ostrakon* (hence "ostracize").

† Around 60 percent of English words come from Greek or Latin, with about an even split between the two. A Greek professor named Xenophon Zolotas delivered two entire speeches in English using only Greek-derived words that could be understood by speakers of both languages.

Most of the time they competed with or fought each other, but in times of crisis they could band together to stand against a common threat. The most menacing came from the East. By proportional population, the Persian Empire was the largest in history, at its height containing almost half of all people in the world across an area stretching from the Mediterranean to India. Although often presented as the despotic counterpart to the glory of ancient Greece, Persia was actually a highly tolerant and cosmopolitan society. It was more like a multiethnic confederation of states than a country. It built roads and bridges and established efficient civil services. To communicate across their lands, the Persians established a courier system whose efficiency (as recounted by Herodotus) inspired the unofficial United States Postal Service motto ("Neither snow nor rain nor heat . . .").

A Persian conquest of Greece may not have been the end of Western civilization. So long as citizens paid their taxes and provided troops for the military, life under the Persians went on almost as usual. And yet, a Persian conquest might well have strangled at birth the golden age that was to follow Greek victory. By 500 BCE, Persia had already conquered the Greek colonies of Asia Minor, and, soon, Greek support for rebellions in the occupied cities sparked conflict. The first invasion by Darius I in 490 BCE was turned back when the Persians were decisively defeated at the Battle of Marathon.* As revenge, Darius's son Xerxes concentrated the largest military force up to that time, numbering one hundred thousand soldiers from across the empire.† This invasion of 480 BCE featured some of the most dramatic moments in military history, including a rearguard action by a contingent of three hundred Spartans who

---

* A triumph for Athens, which prevailed without having to rely on the disciplined but tardy Spartan army, the battle is now remembered for the Marathon run. The story (likely apocryphal) is that the Greek messenger Pheidippides ran the 26.2 miles to Athens, declared victory, and then collapsed dead from exhaustion.

† The exact number is unclear. Herodotus gives a figure of over a million, but this is almost certainly an extreme exaggeration. Modern historians put the number somewhere between sixty thousand and a few hundred thousand. Either way, the army resembled a United Nations assembly, representing dozens of languages and cultures from across the empire, including Persians, Assyrians, Phoenicians, Babylonians, Egyptians, Libyans, Jews, Macedonians, Greeks, Circassians, Afghanis, Indians, and many more.

held a narrow pass at Thermopylae for four days against a Persian host dozens of times its size, and perished in the process.

The Persians eventually broke through and burned Athens (you can still dig up the layer of ash), but the extra time allowed its population to evacuate. The Greeks took refuge in the Peloponnese, a peninsula only accessible via a narrow corridor a few miles wide. Xerxes sent in the navy to outflank them, but in the shallow waters of Salamis Bay the Persian fleet stumbled into a Greek trap and was routed. By almost every measure, Salamis was one of the largest naval battles of all time,* but it was, in fact, mostly a battle of Greeks against Greeks: the Persian fleet was primarily composed of ships from the Asia Minor city-states, supported by strong contingents of Egyptians and Phoenicians. The disaster at Salamis ended the Persian offensive, but it took the equally important but far less famous Battle of Plataea the next year in 479 BCE to end Persia's occupation of Greece.

Victory over Persia left the Greeks unchallenged in the Mediterranean, but soon they were fighting among themselves. After fifty years of unprecedented expansion, the Peloponnesian War (431–404 BCE) pitted Athens and its empire against a league of Greek states led by Sparta. Sparta was stronger on land and soon besieged Athens, but the city was able to hold out for a time by importing food from its overseas colonies. Eventually, Athens was defeated after a disastrous campaign in Sicily, and by starvation and disease in the city. Some suggested burning Athens, but Sparta refused to obliterate a former ally that had fought so bravely in Greece's time of need.† Greek society persisted, but the war per-

---

* Number of ships (over three hundred Greek and twice as many Persian), sailors (two hundred thousand or more), and ships sunk (more than three hundred). Other contenders for this title are the Roman-Carthaginian Battle of Ecnomus (256 BCE), the Christian-Ottoman Battle of Lepanto (1571), and the twentieth-century battles of Jutland (1916) and Leyte Gulf (1944). Supposedly, Xerxes watched the battle from a lawn chair on a hill overlooking the bay.

† In an Athenian comedic drama set in the late stages of the war, the eponymous Lysistrata persuades the women of Greece to withhold sexual privileges from their husbands until they stop fighting. Although there is no evidence this actually happened, something similar has been successfully organized in recent years by women in Liberia.

manently weakened the southern cities and shifted the initiative north, where a new power was rising.

Macedonia was a rural backwater to the Greeks of the peninsula, but it spawned a father-and-son pair of military geniuses who would unite the Greeks and then go on to conquer most of the known world. The father, Philip II, came to the throne in 359 BCE and concentrated on reorganizing the army. Greeks had fought for centuries using spearmen marching in formation, but Philip equipped his infantry with much longer spears, up to twenty feet in length, and drilled them into an elite force. As a formation advanced, interlocking rows of spears provided mutual support and impaled everything in their way. Through a series of wars and alliances, Philip extended his Macedonian domains to encompass the whole of Greece by 336 BCE.* Philip's son, who inherited the throne at the age of twenty, would outdo his father to become perhaps the greatest military genius in history and certainly the best-traveled military adventurer of the ancient world.

By the time he was thirty, Alexander the Great had forged one of history's largest empires, stretching from Greece to India. Like the Greek sailors in the *Odyssey* (a copy of which he always kept by his side), he couldn't stop moving. What drove Alexander? His military campaign against the Persians was initially envisioned as a limited effort to free the Greeks in Asia Minor who still lived under Persian rule, but early successes built momentum, and eventually obsession. It was never a thirst for revenge—rather, it was Alexander's undying fascination with seeing and understanding the world. His campaign was a discovery of lands the Greeks had never heard of, many of which seem remote even today. Driving Alexander was a desire to reach the "ends of the world and the Great Outer Sea."

Although the empire Alexander built didn't long outlast him, it had a profound impact on the world, connecting East and West for the first time. Before, people might occasionally travel to neighboring kingdoms, but they'd rarely venture beyond. Alexander

* Philip supposedly threatened Sparta, "If I enter your lands, I will destroy your farms, slay your people, and raze your city." Sparta's response: "If." Spartans were known for pithy remarks, which is reflected in the English word "laconic," from "Laconia," the region of Greece where Sparta lay.

brought not only an army but an entire civilization to Asia and returned to Greece knowledge from distant lands. Greek philosophers, scientists, and administrators followed in his wake in a massive cultural cross-pollination. Even Alexander's personal style inspired emulation. Before Alexander, most men wore beards. Alexander went clean-shaven, and his preference was picked up by subjects across his empire and copied by the Romans. It's not an exaggeration to say that Alexander made a clean shave manly.*

Alexander's ten-year campaign (334–324 BCE) broke the power of Persia in a series of decisive battles, but he didn't so much destroy the Persian Empire as absorb it. With each victory, he gained credibility at the expense of his counterpart Darius III. Since the Persian Empire was a multiethnic coalition, most of whose citizens weren't Persian, local rulers found it expedient to switch sides and pledge their loyalty to Alexander instead. His military adventures took Alexander through Turkey, down the coast of Syria to Egypt, across the desert into Mesopotamia and Persia itself, through Central Asia, and on to Afghanistan and India. His boldness, bordering on rashness, is illustrated by a (probably apocryphal) story. At the town of Gordium in Asia Minor was a knot of rope so tangled as to be impossible to untie. Prophesy held that it could only be undone by the future "ruler of Asia." A man of action, Alexander decided that it didn't matter how he untied the knot, so he hacked it apart with his sword.

Everything about Alexander's adventure was improvised. He liberated Egypt from Persian rule, was received as a god, and founded Alexandria with its magnificent library. Then he chased Darius across the Middle East into Babylon, and then to the twin Persian capitals of Susa and Persepolis, where he captured the state treasury. He stayed in Persepolis for five months but accidentally burned down the palace in a bout of heavy drinking during one of his notoriously wild parties. Meanwhile, a Persian governor murdered Darius and rose in rebellion, so Alexander set out to quell

---

* The only surviving image of Alexander is a Roman mosaic in the ruins of Pompeii that depicts him as a handsome young man, but we also know from accounts that he went clean-shaven.

the new threat. This turned into a grand tour of Asia, covering many thousands of miles into new and distant lands. Along the way, he founded a series of cities, all named after himself (no one could accuse him of excessive modesty).* He married the princess Roxana from Bactria (in Afghanistan) to cement relations with his new subjects and then turned his attention to India.

Crossing the Indus River in 326 BCE, the Greeks defeated a large Indian army in the epic Battle of the Hydaspes, which included over a hundred war elephants. After the battle, Alexander founded two cities in India that survive to this day, naming one after his horse Bucephalus. Alexander wanted to press on to China, but his soldiers anticipated an endless series of battles ahead and refused to go on. They'd been away for nearly a decade and longed for their homeland. Alexander tried to persuade them to continue, but eventually he agreed to head home by following the Indus south to the sea, where he commissioned his admiral Nearchus to explore the Indian coast and Persian Gulf. Some of the soldiers sailed with the fleet back to Persia, but most endured a grueling march through the Gedrosian Desert, where many perished.

Alexander himself would never return to Greece. He died in Babylon in 323 BCE at the age of thirty-two following a bout of heavy drinking shortly after his return from India. It's unclear if he died of illness or if he was poisoned, but his sudden death created a power vacuum that couldn't be filled. His son and brother were soon murdered, and a chaotic struggle ensued. Eventually, three power blocs solidified in Greece, Egypt, and Babylon, each controlled by one of Alexander's generals. Although the empire had fragmented, its successor states maintained the connection between East and West for centuries. The Greco-Bactrian and Indo-Greek kingdoms in Afghanistan and India developed a hybridized Greek culture, sent the first emissaries to China, and founded the Silk Road as a conduit of goods and ideas between Europe and China.

This first unification of East and West initiated a cultural

---

* Cities founded by Alexander include Kandahar in Afghanistan (it was later renamed) and Alexandria Eschate ("the Farthest") on the border of China.

exchange that was to shape the world. Greek ideas diffused into Asia, and Asian ideas were brought home to the Mediterranean world, influencing art, literature, architecture, music, mathematics, and science.* Greek became the language of scholarship in the West for a thousand years. Biblical names are often Greek because, though Jesus and his disciples spoke Aramaic, Greek was the contemporary universal language. Like Hollywood today, Greece radiated cultural influence throughout the known world, and ambitious parents gave their children Greek names. Even the Roman Empire was half-Greek and acted more as a conduit for Greek ideas than an originator in its own right. Roman scholars spoke Greek, read in Greek, and hired Greek tutors for their children. Many Roman emperors were actually Greek, and later, as power shifted to Constantinople, they *all* were.

The period that followed Alexander's travels was the brightest flowering of scientific discovery until at least the Renaissance. This was the time of the Ptolemies of Egypt, patrons of the Great Library of Alexandria, who levied tolls of not money but knowledge. Ships docked in the harbor would be searched, and any scrolls were copied to be added to the library's collection. Virtually every scholarly field was invented or systematized in these years. Hipparchus invented trigonometry, predicted eclipses, and cataloged celestial bodies. He proposed that stars were born and eventually perished, giving the universe a definitive start and end.† Dionysius of Thrace invented grammar, categorizing the components of speech (nouns, verbs, adjectives). Herophilus mapped the body's digestive system, nerves, veins, and arteries, deducing that the heart is a pump and that the brain is the center of consciousness. Archimedes derived the value of pi and invented parabolic mirrors, screw pumps, pulleys, catapults, and other mechanical devices.

* Many Greeks converted to Buddhism and may have injected ideas into the religion, including divine heroes, incense burning, and gifts of flowers. Some Buddha sculptures also seem to be based on Alexander. Greek astronomical instruments have been found in Afghanistan, and the Greek celestial concepts trickled into Indian astronomical texts.

† A transient universe is also a feature of Hindu cosmology, but it's unclear if Hipparchus was influenced by Greek contact with India.

Some Greek inventions sound downright modern. Aristotle tells of a submarine used to spy on enemy harbors. Apollonius of Perga developed the mathematics of ellipses, parabolas, and hyperbolas, curves that govern the trajectories of projectiles and interplanetary spacecraft. Heron of Alexandria invented gear trains, mechanical robots, automatic doors, medical syringes, steam engines,* and vending machines that dispensed rations of holy water for a coin. In 1902, a device known as the Antikythera mechanism was recovered from a two-thousand-year-old Greek shipwreck. The first known analog computer, it could predict the position of astronomical objects decades in advance using thirty-seven mechanical gears. History has no predefined path, and, had these developments continued unabated, it's possible to imagine an alternative course that would have taken Greek cosmonauts† to the Moon by July 20 of the year 969, a full thousand years before Apollo.

Sadly, it was not to be. Greek society was ultimately encumbered by its high degree of stratification, which vested power in only a small minority. All citizens of Athens could vote, but citizenship was restricted to free landholding males, who accounted for a mere 10 percent of the population. With a third of the population living as slaves, economic and social segregation restricted creative participation to a tiny class. There were shades of racism, too. Egypt was ruled by a Greek king and educated class, while the vast populace saw meager prospects for advancement. The world was divided into Greeks and everyone else. Outsiders were called "barbarians," an ethnic slur directed at people speaking a language other than Greek, who were said to utter an incomprehensible string of noises that sounded like "bar-bar-bar." There are, unfortunately, too many parallels with our own civilization.

But some Greeks actively sought out the barbarians. Around the same time that Alexander was marching across Asia on his mission of conquest, another very different Greek traveled to the

---

* The Greeks invented both the steam engine and railroad but never combined them. The railway at Diolkos in Corinth used pack animals to haul goods on tracks. They were so close to having trains!

† "Cosmos," used in the Russian language, comes from Greek. In fact, the Russian language and alphabet are even more influenced by Greek than ours.

other side of the world on a mission of pure curiosity. His travels would take him across Northern Europe, around the isle of Britain, through the lands of the Germanic tribes, and all the way to the Arctic Land of the Midnight Sun—a phenomenon that he was first to describe. Pytheas "the Greek" was born within a few years of Alexander in the Greek colony of Massalia (now Marseilles, France). For fifteen years, he explored lands to the north that were previously only vaguely known to the Mediterranean world. Upon his return, Pytheas wrote at least eighteen popular works that formed the basis of Northern European geography for centuries to come.*

This was perhaps the first true scientific expedition. Pytheas precisely recorded distances as he went, drew detailed maps, and described natural phenomena in great detail. By measuring the angle of the Sun, he was able to estimate how far north he'd gone.† The surviving descriptions of his travel aren't enough to reconstruct his entire journey, but somehow he arrived in Britain, either by walking overland along trade routes through France or by sailing on a Greek or Carthaginian ship into the Atlantic. We tend to think of Northern Europe at that time as the Greeks would have—a dark land populated by primitive barbarians—but by the time of Pytheas, civilization was creeping in. The Celts and Britons lived in settled agricultural communities and possessed sufficient technology to build sturdy seagoing ships, maintaining trade networks throughout the region. Thus, it may have been a simple matter for Pytheas to book passage on a ship with a friendly crew who were willing to indulge this exotic foreigner's curiosity.

Pytheas lived among the people of southern Britain and found guides there to take him farther afield. Britons had been in con-

---

* Sadly, none have survived, and we only have secondhand accounts from later Greek and Roman scholars.

† This computation of latitude has been used by sailors throughout history. You can do it yourself by pointing one arm at the North Star and another at the horizon. The angle between your arms is your latitude (in the Northern Hemisphere), and each degree of latitude is around sixty-nine miles apart. Computing longitude is another matter entirely, and this problem wasn't solved until an English clockmaker finally came up with a practical method of doing this in 1735.

tact with Carthaginian merchants for a long time, so Carthaginian may have been a common language Pytheas could use to communicate. He describes the inhabitants as simple, hospitable people who enjoyed life, and especially recommends the local grain and honey beverages (beer and mead). Living in simple thatched cottages, these Britons baked bread using grain stored in cellars. Ruled by chieftains, they generally lived in peace. When they did go to war, they fought using chariots (also described later by the Romans), which had long since lost favor in the Mediterranean. Pytheas's British guides brought him up the west coast of Britain to the northern tip of Scotland, which he describes as a frosty land roamed by bears. (This seems strange since there are no bears in Britain today, but evidently there were back then.*)

Pytheas heard tales of the island of "Iern" but apparently never set foot in Ireland. He relates that the Irish were cannibals, but this may have been more British propaganda than fact. In Scotland, he tells of the blue dye locals would use to paint their faces, a customary practice of Scottish warriors from Roman times until the Middle Ages.† In the Orkneys, Pytheas describes ocean tides that rise and fall up to 80 cubits (120 feet)—the real number is no more than 50 feet but still massive compared to that of the Mediterranean, which has a tidal range of mere inches. Here Pytheas also connects for the first time the regular cycles of tides with phases of the Moon. Although a formal description of gravity would wait another two millennia for the arrival of Newton, Pytheas was already applying the scientific method to connect theory with real-world observation.

From Scotland, Pytheas relates that he sailed six days north "as far as the ends of the world" to a place called "Ultima Thule." Greenland, Iceland, Norway, and the Faroe Islands between Scotland and Iceland are all possible candidates. In some accounts, he mentions encountering people, which makes Iceland and Greenland

---

* The association of northern climes with bears persists, with "Arctic" coming from the Greek "*arctos*," for "bear." (The scientific name for a brown bear is *Ursus arctos*, meaning "Bear bear" in a combination of Latin and Greek.)

† Hence the blue face paint in *Braveheart*. The Romans called these people "Picts," meaning "painted."

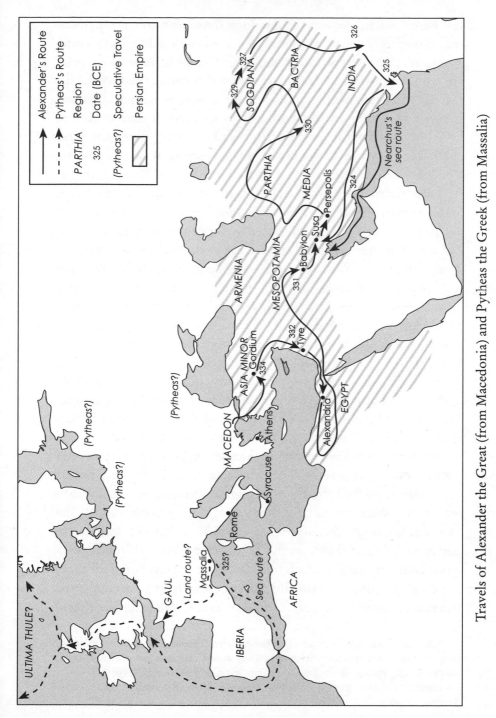

Travels of Alexander the Great (from Macedonia) and Pytheas the Greek (from Massalia)

Legend:

- Alexander's Route
- Pytheas's Route
- PARTHIA   Region
- 325   Date (BCE)
- (Pytheas?)   Speculative Travel
- Persian Empire

less likely, since these weren't settled yet. At Ultima Thule, Pytheas describes a land where the Sun never sets in summer, where winter is shrouded in perpetual darkness, and where the sea congeals into ice that "could not be traversed by foot or boat." Although the Greeks would have been familiar with frost, mountain glaciers, and perhaps an occasional snowfall, a frozen land perpetually covered by ice was difficult to imagine, and contemporary observers were skeptical. However, as we now know, these are accurate observations of the Arctic, suggesting that Pytheas did in fact witness them.

Later travels took Pytheas through Eastern Europe, following amber trade routes that had been operating for centuries. Known as "*elektron*" to the Greeks, amber was purported to have mystical healing properties due to its ability to acquire a static charge — hence "electricity." This journey would make Pytheas the first known person from the Mediterranean world to visit present-day Denmark, Germany, and possibly Poland, meeting tribes such as the "Teutones" along the way. It's unclear if this eastern voyage was part of the same trip or a second one, because he also describes Ukraine and the Don region of southern Russia, which would have been easier to reach from the Greek colonies of the Black Sea. Either way, Pytheas was one of the most widely traveled citizens of the ancient world and one of the first to voyage for the purpose of pure scientific discovery.

Within a hundred years of Alexander and Pytheas, a new power began to rise in the Mediterranean world. With a legendary founding in 753 BCE by twin brothers Romulus and Remus (who were supposedly raised by a she-wolf*), Rome overthrew its last king exactly a year before Athens, in 509 BCE. But while Athens played a central role in the flourishing Greek culture, Rome would remain at the edge of civilization for centuries. An observer would hardly imagine that this tiny city would grow to dominate the entire Mediterranean, but its inhabitants possessed one critical

---

* There is even a theory that the name Rome comes from "*ruma*," referring to the mammary gland of their wolf mother in old Etruscan, which must surely be one of the strangest city name origins ever. *Star Trek* fans will know that Romulus and Remus are the twin home planets of the Romulans, who also use Roman names, ranks, and titles.

characteristic: curiosity. Rome was open to new ideas and quick to adopt them. If an enemy used better weapons or tactics, Rome would copy them. If they found a better means of administration, they'd try it. This openness allowed Rome to control an enormous empire with minimum effort: as long as provincials paid their taxes and stayed in line, Rome rarely interfered.

A better name for the Roman Empire would be the Greco-Roman Empire. Rome conquered Greece in 146 BCE, but to a large degree it was the conquerors who were themselves assimilated. Greek culture was so dominant that even at its height, the empire was a Latin-Greek fusion, with Rome providing the administration and Greece providing the scholarship. Greek remained the dominant language of education and was widely spoken even by Romans from Italy. A similar cultural assimilation recurred later when "the barbarians" overran Rome and, in essence, became Roman. The spread of Greco-Roman culture into French, British, Spanish, and German lands is why it still persists in Western Europe and, by extension, North America. It's why the US Capitol is built in classical style and is named after the ancient Roman administrative center atop the Capitoline Hill. It's why the Roman eagle has been chosen as the symbol of dozens of countries. The months on our calendars have Roman names.* This book is printed in Rome's Latin alphabet, in a fusion Latin-Greek-Germanic language.

Rome's expansion brought it into conflict with Carthage, a Phoenician city-state that controlled the southern Mediterranean. Rome defeated Carthage in the Punic Wars ("Punic" is derived from "Poeni," the Roman name for Carthage), despite Hannibal's heroic efforts to famously march elephants across the Alps from Spain. Rome won because it was able to draw on manpower reserves from its citizenry† and because it was willing to adapt.

---

* The days of the week also derive from Rome, but some have Anglo-Saxon gods substituted in (e.g., Roman Tuesday was *dies Martis*, "the day of Mars"; the Anglo-Saxons substituted their own god of war, Tiw, and their own word for "day," and got *tiwesdaeg*).

† Mary Beard in *SPQR* estimated that during the late republic, more than half of male Romans had served in the army overseas, in Greece, Spain, Africa, and the Middle East. This might make Romans the best-traveled people in history.

Rome was a terrestrial power, but it learned shipbuilding by studying a Carthaginian shipwreck off the coast of Italy. Just nine years later, Rome had built a navy and defeated a Carthaginian fleet of three hundred ships off Cape Ecnomus in Sicily. Two captured Carthaginian ships were paraded through Rome, signaling its rise as a naval power. Eventually, Rome controlled every inch of Mediterranean coastline, and the sea was dubbed "Mare Nostrum" ("Our Sea"), a name that Mussolini tried to revive in the 1930s.

By the time of Augustus Caesar at the turn of the millennium, the Roman Empire spanned 1.8 million square miles (almost two-thirds the size of Australia or the contiguous United States). It was connected by an efficient trade network, with roads and bridges so expertly built that many are still used. There were over fifty thousand miles of paved Roman roads (enough to stretch twenty times across the United States). The main highways of Europe are laid out based on the old Roman ones. This was a multiethnic empire, knitting together people from dozens of cultures. Rome itself was highly cosmopolitan (a Greek word); it was the first city to reach a million inhabitants, a level not surpassed until nineteenth-century London. It would not have been unusual to find Africans and Middle Easterners living and trading in Rome, and not impossible to find travelers from as far away as India and China. Anyone could become a Roman citizen, and many foreigners did, some finding employment as prominent playwrights, philosophers, generals, politicians, and even emperors.*

With their overseas connections, the Romans possessed vastly better knowledge of the world than any Europeans until the Age of Exploration more than a thousand years later. After acquiring Egypt in 30 BCE during the affair of Cleopatra and Mark Antony, Rome developed trade routes across the Red Sea to India. According to the Roman geographer Strabo, a fleet of over a hundred ships made an annual voyage to India and Southeast Asia. Like the later Portuguese, the Romans timed their voyages to coincide

* By the late empire, more Roman senators came from outside Italy than within. In 212, Emperor Caracalla (of African origin) granted citizenship to all free people within the empire's borders, the largest citizenship grant in history: twenty million at a stroke.

with the monsoon winds. Exotic African and Indian animals such as lions, tigers, rhinoceroses, elephants, and cobras were imported for circus and gladiator shows. Fashionable Roman women wore Indian shawls, pearls, and scarves dyed in indigo.

The kitchens of wealthy Romans were stocked with Indian sugar, herbs, and spices. Pepper was especially prized in Roman cooking. Pliny the Elder complains in his *Natural History* that "black pepper is fifteen denarii per pound, while white pepper is seven," and that "there is no year in which India does not drain the Roman Empire of fifty million sestercii." At a rough equivalence of a few US dollars per sestertii or twelve per denarii, this translates to over a hundred dollars per pound of pepper, and a yearly trade of more than a hundred million dollars.* This would mean six hundred tons of pepper imported each year from India, unmatched until Portugal and Venice restored the old trade routes in the mid-1500s. More evidence of connection comes from Roman coins found all over India.†

There was also a tenuous link between Rome and China. Emperor Augustus is recorded to have sent an envoy to "Seres," which was probably China, although the term was used for the entire Far East.‡ The first definitively known Roman embassy is recorded in Chinese sources at the time of Marcus Aurelius, around 160 CE. The envoys probably traveled by sea, entering China from Vietnam, and reportedly brought gifts of ivory, rhino horns, and tortoiseshells. Several Roman private citizens also visited China. A traveler named Maës Titianus walked along the Silk Road at least as far as western China around 75 CE. Around 220 CE, a Roman merchant visited the imperial court of the Wu Kingdom (one of the Chinese political units at the time), where

---

* In case you were wondering about Roman money, one gold aureus = 2 gold quinarii = 25 silver denarii = 50 silver quinarii = 100 bronze sestertii = 200 bronze dupondii = 400 copper asses = 800 copper semises = 1,600 copper quadrantes.

† St. Thomas supposedly visited India in 52 CE, and Christians and Jews established communities there during Roman times. Individual missionaries probably made it as far as China.

‡ Roman maps separated the land of silk (Serica) at the end of the overland Silk Road from the land of the Qin (Sinae), which could be reached by sea, even though both were the same place.

the emperor Sun Quan greeted him, asked him about Rome, and organized a sea expedition to return him home.

These experiences are strikingly similar to Marco Polo's a thousand years later, emphasizing that later European expeditions to Asia weren't so much novel discoveries as rediscoveries of the world the Romans had known. It seems clear that there was sustained, if sporadic, contact between China and Rome, and even some trade. For their part, the Chinese sent some of their own envoys to Rome, which they called "Da Qin," meaning "Big Empire,"* imagining it as an equivalent counterweight to China at the far western end of Eurasia, effectively keeping the world in balance. The most detailed surviving account is that of Gan Ying, who traveled west in 97 CE but may have only gotten as far as the Persian Gulf at the edge of the Roman world. Possibly based on secondary sources, Gan Ying's account provides a fanciful description of Roman governance, noting that Romans didn't have permanent leaders but instead selected the best candidate for the job, who could easily be deposed when a better ruler came along.[†]

There's another way Romans could have reached China. After their defeat at the Battle of Carrhae in 53 BCE, at least ten thousand Roman soldiers were captured by the Parthians, who ruled an empire stretching from Mesopotamia to Afghanistan. Some of these prisoners may have been incorporated into the Parthian army and transferred east, eventually to face the Chinese seventeen years later in the Battle of Zhizhi, where Chinese sources reference strange warriors in "fish-scale formation," reminiscent of the Roman "*testudo*" (tortoise) of interlocking shields. According to local legend, these warriors were captured by the Chinese and settled in Liqian, where their "descendants" claim Roman heritage today.[‡] Since this proposition is at the end of a convoluted

* The direct translation is "Big China," since there is no distinction between "China" and "empire" in Chinese.

† This was probably his interpretation of Roman elections for the Senate, which actually outlived the empire, convening until at least 603 in the west and 1204 in the Byzantine east.

‡ DNA tests conducted on locals have been mixed but haven't conclusively demonstrated a Roman connection.

chain of events, an alternative explanation may be that the warriors were simply descendants of Alexander's Greeks, who, after all, ruled kingdoms in Central Asia for centuries (the last disappeared from the Indo-Chinese border more than sixty years after those events). Whatever the particulars, it's clear that people of the classical world traveled vast distances, and thereby acquired a better understanding of the world than anyone on our planet for the next thousand years.

PART II

REDISCOVERING
THE WORLD

6 | BARBARIANS FROM THE NORTH

No single event marks the fall of the Roman Empire. Key mile-stones in its decline from a height in 117 CE include the adoption of Christianity by Constantine in 312; the permanent split into eastern and western halves in 395; the sack of Rome by Visigoths and Vandals in 410 and 455, respectively; and the overthrow of the last emperor on August 28 of the year 476. By this time, the empire had fractured into at least twenty successor states, including the Eastern Roman Empire, which was to persist for another thou-sand years.* But in many ways, the Roman Empire never ended. The successor kingdoms that took its place, such as those of the Visigoths in Spain, Ostrogoths in Italy, Vandals in North Africa, and Franks in Western Europe, didn't so much conquer the empire as become it, with their adoption of Roman language, culture, and religion. Europe became a patchwork of small independent states, each tracing its heritage back to Rome.

We tend to think of "the barbarians" as violent nomads bent on conquest, but they were mostly just simple pastoral folk, many of whom had lived within the borders of the empire for a long time. What drove most of them was simply a desire to be Roman. Many

---

* This "Byzantine Empire" still thought of itself as Roman in every way, except that it spoke Greek. It even managed to retake Italy and reincorporate Rome into the empire for a time. Early Russian rulers married into this line of emperors, thus claiming the title "tsar" ("Caesar"). Even Germany had a tenuous claim through the Holy Roman Empire, thus the term "kaiser."

had served as hired mercenaries in the Roman army. Some fought against the empire and some for it—Theodoric of the Visigoths, for example, forged an alliance with Rome to stop Attila's pillaging Huns.* Nor was there as much of a technological disparity between Rome and "the barbarians" as we might imagine. Over the centuries technology had diffused far ahead of the empire's borders. When Julius Caesar invaded Britain in 54 BCE, he found an Iron Age culture with a population numbering in the low millions. They already used coins and writing and were already trading with the Mediterranean world.

Northern Europeans had built sturdy seagoing vessels for centuries. A Bronze Age ship dating to hundreds of years before the Romans' arrival was unearthed in Dover in 1992. Built from oaken panels and thirty feet long, the vessel supported a crew of a dozen. In 56 BCE, Julius Caesar fought a naval battle off the French Atlantic coast against the Gaulish Veneti tribe in a contest that engaged a hundred ships on each side. He described the Veneti ships as being far more seaworthy than the Roman ones, being constructed of solid oak planks fixed by iron nails "as thick as a thumb," capable of withstanding the harsh conditions of the Atlantic. The thick hulls of the Veneti vessels made them resistant to ramming (the most common Mediterranean naval tactic), and their maneuverability made them difficult to board. All the while, the Veneti ships showered the Roman galleys with arrows from tall masts. This very real naval threat posed by supposed "barbarians" led Rome throughout its rule to maintain an anti-piracy fleet in the North Sea.

Ireland was never conquered by Rome, but the Romans enjoyed an ongoing commercial relationship with the island they called "Hibernia." Roman merchants traveled there, and, judging from archaeological finds of Roman equipment, there were probably a few small military expeditions. While the empire was collapsing in the early fifth century, a Roman citizen named Patricius

---

* For coming to Rome's aid against Attila at the Battle of the Catalaunian Plains, where he was killed, Theodoric was revered in Europe as a savior of civilization, providing inspiration for J. R. R. Tolkien's Théoden of Rohan in *The Lord of the Rings*.

was kidnapped by Irish pirates and brought back to Ireland as a slave. After a few years, he escaped and returned to Britain, where he became a cleric, whereupon the future St. Patrick returned to Ireland as a missionary to spread the teachings of Christianity. The Irish took to the new religion with zeal, and Irish monasteries played a major role in preserving ancient Greek and Roman manuscripts, despite their being written by pagans. One of these early Irish monks gives us some clues about Irish travels of this period.

St. Brendan was probably born a few years after Patrick, somewhere around the end of the fifth century. There are several versions of his *Voyage of St. Brendan the Abbot*,* but most have him setting off in the early sixth century with a crew of sixteen in a traditional seagoing Irish currach built of wood and ox hide. This was a spiritual journey for Brendan and his companions, who sailed into the unknown to literally get lost, thereby demonstrating their convictions.† Along the way, they found a dozen islands, some inhabited by people, others by sheep, and some by mythical beasts such as griffins and birds of paradise. Most of the tales are fanciful or symbolic, but a few interesting tidbits may be based on real observations. "Demons throwing rocks from mountains of fire" and "crystal pillars jutting out of a congealed sea" sound an awful lot like volcanoes and icebergs. A "submerging island" could refer to a diving whale, drawing parallels with other classic seafaring tales from Jonah to Sinbad the Sailor.

It's unclear where St. Brendan actually visited, if anywhere. Iceland is a prime candidate, and Greenland is also possible. Some have suggested the Azores or Canary Islands, which lie to the south but are easily reached from Ireland. Whether St. Brendan got anywhere or not, people assumed his journeys were real, so medieval European cartographers adorned maps of the Atlantic with islands based on his descriptions. Columbus sailed on his

---

* The account, which may have been written after his death or be based on a fictional person, was translated into many languages and became a bestseller in Middle Ages Europe. Columbus is thought to have owned and been heavily influenced by it.

† Celtic tradition considers going west across the ocean to be heavenly, hence the "into the west" theme used for the elves in Tolkien's *The Lord of the Rings*.

voyage with a map that showed an "Island of St. Brendan" out in the Atlantic at roughly the distance of North America. In fact, it was widely accepted by Europeans that there were unknown lands out in the Atlantic—the only question was whether these were mere islands or part of the Asian mainland.

Did St. Brendan visit America? There is no evidence that he or any other early Irish navigator made it that far, but minimalistic monks wouldn't tend to leave much archaeological evidence. It's certainly possible. Irish seafaring monks visited Iceland around this time, and from there it's less than two hundred miles to the coast of Greenland. From there, it's only another two hundred miles to Canada's Baffin Island, or five hundred miles to Labrador. Today, a St. Brendan's Society builds traditional boats and champions the idea that the early Irish reached America. In 1976, an adventurer named Tim Severin built a thirty-six-foot-long currach using only technologies available at the beginning of the sixth century to prove this could have happened. In his boat, dubbed the *Brendan* (described as a "giant leather banana"), Severin sailed from the Faroe Islands near Scotland all the way to Newfoundland.*

There was one group of early northern explorers who did conclusively reach America. These were the Vikings, who were suddenly unleashed on an unsuspecting Europe on June 8, 793. That day, a Norse raiding party landed on the shores of Northern England to slaughter the community of monks at the monastery of Lindisfarne and carry off everything of value. This violence perpetrated on a peaceful religious community by mysterious warriors sent shockwaves through Europe as far as Charlemagne's court in Aachen. To commemorate the slain, the "Domesday Stone" was erected, depicting, on one side, ferocious warriors brandishing swords and battle-axes, and on the other, the end of the world. Medieval Europeans were totally unprepared for the Viking raids to follow and could find no explanation except "the wrath of God." It was the start of the Viking Age.

---

* Another legend has Christian bishops from Spain leading their flocks across the Atlantic to escape the Moorish invasion. They settled on "Antillia," an island that appeared on medieval maps and lent its name to the Caribbean Antilles.

Driven by fierce competition for land back home, these sea-faring Scandinavian warriors would terrorize the continent for two centuries, until they eventually adopted Christianity, inter-mixed with locals, and settled down. Starting from their homes in Denmark, Norway, and Sweden, they raided, traded, and set-tled across a wide swath of Europe from Russia to Iceland. They ranged south into the Mediterranean as far as Sicily and North Africa, and through the rivers of Russia to the Black Sea. They founded communities and kingdoms that still thrive today. It was a cultural exchange on a greater scale than any since classical times, and it brought Europeans across the Atlantic for the first time.

More than any other event, the attack on Lindisfarne shaped perceptions of the Vikings. It wasn't until the 1890s that schol-ars outside Scandinavia seriously reassessed the Vikings in terms of their technology, seamanship, and exploratory achievement.* Like the Greeks, the Vikings prized individuality and curiosity. Boldness and bravery were woven into their myths: upon death, a Viking would travel to Valhalla, where there'd be constant fighting and feasting. Any warrior who was killed would be reincarnated the next day to do it all over again (this doesn't sound far removed from some descriptions of hell). The Vikings prided themselves on their skill in battle; their ability to work themselves into a frenzy of fearlessness has left us with the word "berserk." From the Norse for "bear coat," the word suggests "becoming like a bear."

The Vikings had a pantheon of gods with humanlike character-istics, many of whom were shared with the Angles and Saxons who settled Britain and inserted the gods into our days of the week: Tiw, the god of the sky (Tuesday); Wotan, the leader of the gods (Wednes-day); Thor, the god of thunder (Thursday); Frigga, the goddess of marriage and fertility (Friday).† For all their supposed violence,

---

* "Viking" probably comes from "*vik*" ("ocean inlet"), so that to go "viking" meant sailing into bays and rivers to raid. Not being a unified group, these Scan-dinavian warriors adopted different names, depending on their location. Those who settled in Ireland, for example, called themselves "Ostmen" ("Westmen"), since Ireland is west of Scandinavia.

† Vikings also worshipped more obscure gods, like Ullr, the god of skiing. He's depicted on rune stones in full regalia, ready to hit the slopes.

Vikings lived in a remarkably free society, at least by the standards of the time. Although they maintained a class hierarchy consisting of jarls (nobles), karls (free men), and thralls (slaves),* all classes mixed in society and enjoyed at least some protection under the law. Karls tended to be less wealthy than jarls, but they enjoyed the same rights and freedoms, meaning that the Vikings were one of the first societies with a sizable middle class. Thralls were usually non-Vikings captured in raids. Though they were slaves, thralls lived a life that was far less harsh than that of slaves in later ages, being able to own property and buy their freedom, and were frequently released for good service. Viking women usually married young at the discretion of their fathers, but they had significant power for the time, able to inherit and own property, run a business, take legal action, and even divorce an abusive husband and receive alimony.

Ships were central to Viking culture, with the dead often buried in them with all their possessions, ready for their trip to Valhalla (entire vessels are often unearthed loaded with artifacts). The most famous was the longship, agile and shallow drafted enough to move upriver to conduct raids. Propelled by both sail and oar, some longships were massive: up to 120 feet long, with a crew of almost a hundred. For long voyages or to transport cargo, Vikings used the *knarr*, a broad sailing vessel with a deep keel. *Knarren* could be as big as Columbus's ships—longer and wider than a railway freight car. These were the ships of choice for voyages to the Mediterranean and across the Atlantic. Both longships and *knarren* often mounted a dragon head at their bow, intended to frighten sea monsters, evil spirits, and no doubt human adversaries, too.† Much of our naval terminology has been handed down from the Vikings, including "keel," "raft," and "starboard," which referred to the giant rudder "steering board" on one side of a ship (the other side was later named "port," for the side a ship docks on).

On long journeys, the decks of Viking ships were crammed with

* Hence "enthralled" for "being under someone's power."
† The head was often removable to avoid alarming friendly land spirits, making it something like a skull-and-crossbones pirate flag that could be raised before battle. Many ships were elaborately decorated, some to strike terror into their enemies, others just to showcase the individual personalities of their crews.

people, supplies, and even live animals such as cattle, oxen, sheep, dogs, and horses. Fires on board weren't permitted lest they ignite the ship, hence, food had to be eaten dry, salted, or pickled. This turned out to be a secret weapon, because, unbeknownst to the Vikings, their copious consumption of onions, radishes, and sauerkraut provided vitamin C that protected them from the scurvy that plagued other sailors until the late eighteenth century. The Vikings were expert navigators, sailing by the position of the Sun, Moon, and stars. Like the Polynesians, they also relied on natural cues such as tides, waves, currents, winds, clouds, and birds. Instead of magnetic compasses (which had already been invented in China but hadn't yet made it to Europe), Vikings used a "sun compass" that computed direction based on the angle of the Sun, like a sundial. They may also have used a "sunstone," a transparent mineral that polarized light to help locate the Sun, even in an overcast sky.

Around the year 800, Vikings started venturing beyond their Scandinavian homeland. Their wanderlust began with sporadic raiding, but as time went on, more and more decided to remain overseas. By 850, Viking armies were overwintering in England. By 870, Danish chieftains founded permanent kingdoms there, effectively splitting the country with the Anglo-Saxons after a struggle ended in a stalemate. Even today, most towns and cities in the so-called "Danelaw" of eastern Britain have names of Danish origin. Eventually, England was united under King Canute in 1016, who also ruled Norway and Denmark, only to fall in 1066 to the Normans led by William the Conqueror,* which added French to the great melting pot of Britain. But the Normans themselves were originally Vikings who'd been allowed to settle in France in exchange for agreeing to stop plundering Paris on their frequent raids up the Seine.†

In the east, Vikings sailed across the Baltic to fan out through the interconnected rivers of Russia. Here they encountered peo-

---

* Until his conquest of England, he was known as "William the Bastard." Not as glamorous.

† In the largest raid, in 845, Vikings sacked Paris and demanded a ransom of 5,700 pounds of gold and silver to leave. This raid may have been led by the legendary Ragnar "Shaggy Pants" Lothbrok.

ple called "Slavs," whom they pressed into service to haul boats between rivers (hence "slave"). Viking merchants made fortunes sailing down the Volga and on to Baghdad, one of the richest cities on Earth, whose bazaars were filled with exotic goods from as far away as India and China. Viking chiefs founded kingdoms at strategic points along Russian rivers. One such kingdom, established at Kiev on the Dnieper, became Kievan Rus, which, through a succession of rulers of the Rurik dynasty up to 1598, laid the foundations for modern Russia. Vikings even provided the name for Russia; "Rus" is derived from a term for "oar." In the Byzantine Empire (the continuation of the Eastern Roman Empire), Vikings were known as "Varangians." At first, they raided towns along the Byzantine-controlled Black Sea from their rivers in Russia and Ukraine. This only stopped when the eastern emperor decided to recruit them as mercenaries, whereupon they formed the imperial bodyguard, thenceforth known as the "Varangian Guard."

In the west, Vikings from Norway set out around the year 800 to settle the Shetland and then Faroe Islands. The Shetlands lie just north of Scotland,* and there the Vikings found rich fishing grounds and grazing pastures. Today, the Shetlands are home to a host of diminutive animals, including dwarf sheep prized for their fine wool, ponies no taller than a large dog, miniature "sheltie" sheepdogs, and cattle among the smallest in the world (small livestock have a reduced grazing footprint in the fragile ecosystem, but Shetlanders also compete to breed smaller animals for bragging rights). Farther out, between Scotland and Iceland, the Faroes are rockier, wetter, and cloudier than the Shetlands; indeed, they're among the foggiest places on Earth. The islands get half

---

* The Shetlands are part of Scotland's oil boom but might not be for long. They were given by Denmark to Scotland upon the betrothal of Margaret of Denmark to King James III of Scotland in 1468, but there is a clause that Denmark can demand them back at any time upon payment of a ten-thousand-kroner dowry. Scottish agents supposedly destroyed all records of the deal in 1579, but the current queen of Denmark claims to have a copy in her possession. If true, Denmark could try to repay the dowry, possibly in response to a Scottish vote for independence from the United Kingdom. How much is ten thousand kroner today? With inflation, it might be as much as $3.7 billion—which sounds like a lot but might actually be a bargain given the oil reserves at stake.

as much sunshine and 60 percent more rainy days than famously soggy Seattle. This makes the Faroes more suitable for fishing than agriculture, but they, too, have miniature breeds of sheep, cattle, and ponies—but slightly larger than the Shetland varieties, at least according to officials who measure these sorts of things. The Faroese language is one of the closest modern relatives to the old Viking tongue.

In 841, the Vikings built a trading post at Dublin just down-river from a Gaelic settlement, from which they traded and raided along the coast. Some intermarried with locals, while others sailed on, spurred by Irish tales of lands to the west. According to the sagas, in 868 a Viking named Flóki discovered a large island he called "Snæland" ("Snowland"), where he decided to overwinter. In spring, he found the land (Iceland) warmer than expected, covered in lush green pastures. After sailing on to discover Greenland a further two hundred miles west, Flóki then returned to Norway to report the uninhabited lands ripe for settlement. Starting around 874, a rush of settlers fleeing the harsh rule of the Norwegian king flooded Iceland, landing at Reykjavík ("Smoke Cove"), named for the nearby geothermal steam vents. By the early 900s, all the best land was claimed.

A council of chiefs was set up to regulate disputes,* but since there was no central executive like a king, Icelanders often challenged the rulings, which led to a lot of violent clashes and blood feuds. One of these involved Erik the Red, named for his fiery hair and beard. Erik was banished from Iceland for three years for "some killings" around the year 982. Apparently an argument arose over ownership of some ornate planking beams from Norway that he was saving to build into the ceiling of his house. A man named Thorgest was supposed to look after them until Erik's house was ready. When Erik went to claim them, Thorgest insisted they were his (which they may well have been; the sagas aren't

---

* Formed in 930, this evolved into Iceland's modern parliament, which is reckoned to be the oldest in existence. Originally, the council debated and passed laws, but these weren't written down: all laws were memorized by an elected "lawspeaker" who served as arbitrator. Can you imagine if modern legislation worked like that?

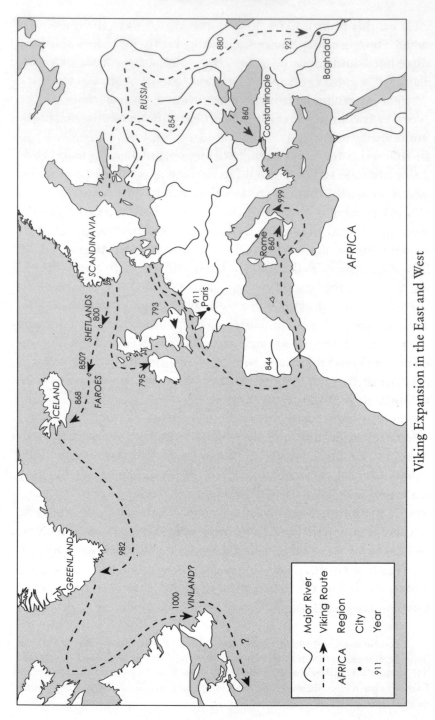

Viking Expansion in the East and West

clear on this point). Later, Erik ripped the planks out of the wall when Thorgest wasn't home and walked off with them. Thorgest came home and gave chase, and in the ensuing fight, Erik ended up slaying Thorgest's son and "a few other men." Yeah, weird.

Feuding ran in the family: Erik's father had been banished from Norway for killing a rival and had fled with his family, including young Erik, to Iceland. Now Erik had to move on, so he decided to follow up on Flóki's report of other lands to the west. Erik spent his exile sailing around the coast of Greenland (the largest island on Earth), finding ice-free fjords and green pastures in the south, not unlike those of Iceland. Upon his return to Iceland, Erik announced the discovery of a verdant land perfect for settlement. In an obvious public relations campaign (if not blatant false advertising), Erik chose the name "Greenland" as a favorable contrast with the barren-sounding "Iceland."* After a recent famine, many Icelanders were already seeking greener pastures (so to speak), so they followed Erik to the new land.

In 986, a fleet of twenty-five ships set out from Iceland with colonists bound for Greenland. They established two settlements (eastern and western) in the two areas best suited for farming. In Greenland (technically part of North America), they built a life around European agriculture, with their cattle, sheep, goats, pigs, chickens, barley, and wheat. They harvested grass for hay to last the long Greenland winter and cut sod to insulate their houses. In summer, Viking hunters sailed up the coast in search of walruses for ivory, seals for meat and blubber, and even live polar bears to export to Europe as royal novelty pets. The Vikings also hunted whales (although they weren't as well equipped for this as the Inuit) and exported narwhal tusks to Scotland, where they were passed off as unicorn horns possessing magical properties. Thus, Scotland, and by extension, Canada and the United Kingdom, acquired fantasy animals emblazoned on their national coat of arms.

---

* Greenland and Iceland were a few degrees warmer back then, thanks to periodic climate cycles. Iceland is one of the youngest landmasses in the world, having formed within the past twenty-four million years at a fissure in the Mid-Atlantic Ridge. Situated right next door, Greenland is one of the oldest landmasses, having formed over 3.5 billion years ago.

At its peak, the Greenland colony supported a population of over five thousand, boasting at least four hundred farms whose sites have been identified by archaeologists. A bishop appointed by Rome presided over the cathedral (more like a large stone church) at the village of Gardar, the most remote outpost of Europe. Jared Diamond in *Collapse* has suggested that the cost of supporting the bishopric, in imported robes, golden goblets, and sacramental wine, was a serious resource drain, symbolic of how obstinately the Greenlanders clung to their European roots. Their refusal to adopt hunting techniques better suited to the climate, including their apparent disinclination to eat fish—the one reliable food source—eventually doomed the colony. But we shouldn't judge them too harshly. The Greenlanders were quite successful, lasting almost five centuries, until the early 1400s, which is longer than people of European descent have lived within the borders of the continental United States.

From the start, there were hints of more undiscovered lands. One of the first settlers, Bjarni Herjólfsson, got lost during his move to Greenland and accidentally sailed too far west. Upon sighting North America (possibly Labrador), he sailed along the strange coast for a few days before heading back to Greenland. He reported seeing thick forests in the new land, but not much else. A few years later, in the year 1000, Erik the Red's son Leif Eriksson was chosen to lead an expedition to follow up on the report. Sailing west, Eriksson and his men first came to a desolate rocky place they named Helluland ("Rock Land"), probably Baffin Island in the Canadian Arctic. Farther south, they landed in a thickly wooded region they called Markland ("Forest Land"), probably Labrador on the Canadian mainland. Finally, after two more days of sailing, they arrived at a beautiful land with green meadows, rivers full of salmon, and plentiful grapes or berries (the Norse used the same word for both). They called this place "Vinland." We're not sure exactly where Vinland is, but probably somewhere between Newfoundland and New England. Since it was late in the year, Leif and his crew built a small settlement, where they spent the winter.

In the 1960s, Norwegian archaeologists Helge and Anne Ingstad discovered the remains of a Norse settlement at L'Anse aux Meadows along the northern tip of Newfoundland. This settlement was large, supporting a population of up to a hundred. It consisted of almost a dozen sod-covered wood-framed buildings, with a main hall one hundred feet long by fifty feet wide (the size of an NBA basketball court). Several smaller workshops and dwellings filled out the settlement, which included a forge with iron slag, a carpentry shop with saws and wood chips, and a boat repair shop where iron rivets were found. The presence of a spindle and needle suggests that women lived there— meaning this was no mere winter camp, but probably an attempt at a permanent settlement. Remains of butternuts found at the site imply that the settlers ventured at least as far south as the northern tip of the tree's range in New Brunswick, six hundred miles away.

The sagas describe other settlements at least a few days' sailing south of Leif's camp. This would put them in Nova Scotia or Maine, or possibly as far as Massachusetts or Long Island. Based on these descriptions, the Viking presence in North America was no mere voyage of exploratory curiosity: it was a sustained attempt to settle a new land. Why did it fail? For one thing, North America was very far away considering the technology of the time. Even Greenland was a frontier outpost at the extreme limit of European influence, constantly in danger of losing touch with the mother continent. There just weren't many settlers in Greenland who were willing or able to settle even farther west. For another, unlike Greenland, continental North America was already populated by people who were better equipped to live there.*

Around 1004, Leif's brother Thorvald sailed with a crew of thirty and spent the winter in North America, where he encountered Native Americans for the first time. Coming across people sleeping under canoes on the beach, Thorvald attacked and killed eight with no apparent explanation or justification. The ninth escaped, and (no surprise) came back with a war party to attack

---

* There were Inuit in Greenland, but they arrived after the Vikings and may have contributed to their demise.

the Viking camp. Thorvald was killed by an arrow in the ensuing struggle. Thorvald's sister Freydís was pregnant, but nevertheless picked up the sword of a man who had been killed and rallied the men to fight. According to the saga, her fierce demeanor frightened off the natives, called "Skraelings" by the Vikings, possibly derived from "*skrækja*" ("shout") on account of their "unrecognizable" language.*

This encounter didn't end Viking attempts at settlement. A few years later, Thorfinn Karlsefni "the Valiant" sailed three ships loaded with supplies and settlers bound for America. They headed south to a place called Straumfjord, possibly in Newfoundland, Nova Scotia, or Maine. There, they initially enjoyed peaceful relations with the natives and managed to exchange some goods. Around this time, Thorfinn's son Snorri† became the first European born in the Americas. Around 1014, another expedition brought additional settlers, but after that, Viking America disappears from the written record. But voyages to America seem to have continued for a few centuries, at least sporadically. A Norwegian penny dating to the late eleventh century found at a Native American site in Maine suggests Vikings were in New England a hundred years after the founding of the Vinland settlement. Other Viking artifacts have been found in the Canadian Arctic and at scattered Inuit sites. In one of the last known references to Viking America, Icelandic records from 1347 mention a crew returning from Markland with a load of timber.

Were there more American settlements, and if so, what became of them? Were they abandoned soon after their founding, or did they last a century or more? Would we even be able to tell from the scant traces of archaeological evidence? Most people used to assume that *all* stories of Vikings in America were sheer fantasy.

---

* The sagas report that Vikings captured two native children and brought them back to Greenland, and there is DNA evidence suggesting at least one Native American woman was captured and brought to Iceland (the first Native American in Europe), where her descendants still live today.

† In contrast to how it sounds in English, the name "Snorri" actually describes a mighty warrior. Later in life, Snorri Thorfinnsson would move to Iceland, where he became a respected political leader.

L'Anse aux Meadows is a relatively recent revelation, and it was only located because the grassy mounds of former buildings stood out on a windswept Newfoundland plain. If there were Viking ruins buried in a forest somewhere along the coast of Maine, we might never know. It's possible that Vikings lived in North America for a long time but evidence has been either erased by more recent European arrivals or just not yet discovered. However, until and unless we find more clues, it's probably safe to assume that Viking settlers didn't inhabit mainland North America for long because they were either unable or unwilling to peacefully coexist with the inhabitants of the continent.

Small groups of Vikings didn't have the advantages of numbers or diseases that allowed later Europeans to overwhelm the native population. It wasn't that contact with a single European would automatically transmit deadly diseases: there had to be an infected person in close proximity to a potential victim. Sporadic contact with a few Vikings simply wasn't enough. The mass epidemics that virtually annihilated Native Americans didn't start spreading until many years after Columbus's arrival, by which point thousands of Europeans were swarming through the Caribbean. Nor did the Vikings enjoy a significant technological superiority. A few iron swords wouldn't have made much difference in small-scale combat, especially when the natives could bring many more warriors to bear.

The point is illustrated by King Philip's War in the 1670s. A coalition of Native Americans led by a chief named Metacomet (he went by "Philip" in English) tried to resist colonial encroachment at around the last time Native Americans could face Europeans on almost equal terms. In the space of a year, more than half of New England's colonial settlements were attacked, and more than a dozen were razed. Several thousand European settlers were killed (around 10 percent of the population), and the colonial economy was ruined. The fact that a fifty-year-old colony with tens of thousands of settlers benefiting from European diseases, more advanced firepower, and native allies had difficulty surviving on the shores of North America should make it clear just how precarious the Viking situation was. Columbus's first settlement on Hispaniola was destroyed in 1493. The first English colony on Roanoke Island was abandoned in 1587. Jamestown and Plymouth barely survived their first years, and probably wouldn't have except for influxes of new arrivals to fill the ranks of the dead. Establishing a permanent presence in a new land is hard.

Even the Viking settlements in Greenland were eventually wiped out, in large part due to cooling climates of the Little Ice Age (1300–1850). Native competition also played a role, judging from Viking combs, utensils, chess pieces, rivets, tools, and other artifacts found in Inuit possession (though these may have been acquired through trading, raiding, or scavenging from abandoned sites). Other factors in the Vikings' decline were ivory competition from emerging trade routes in Africa and, notably, an epidemic that killed around half the population of Iceland around 1400 and may have spread to Greenland (probably a remnant of the Black Death, which had spared Iceland up to that point). A marriage certificate from 1408 is the last written record from Viking Greenland, and a few artifacts have been radiocarbon-dated to as late as 1450. However, it is just possible that some Greenlanders persisted another forty-two years until Columbus's voyage in 1492, which would constitute a continuous European presence in North America dating back to the year 986.

7 | EARLY CONTACTS

At least one Viking settlement in North America has been defin-itively proven. Other Viking settlements seem likely based on extrapolation. Though the sagas were embellished with fantasy elements, the historical events they describe have proven remarka-bly accurate over the years. In addition to traces left of the Vikings, there are many other tantalizing clues of early contacts between the Americas and the Old World that may well have occurred but for which there isn't enough evidence to say for sure. Written records are sporadic or nonexistent, and archaeological clues of small-scale contact might be difficult to find or distinguish from subsequent activities. From a skeptical scientific perspective, we can't include these contacts in the realm of verified fact. However, some seem plausible enough to mention, and, given the vagaries of chance and time, it seems likely that at least some occurred.

One of the most intriguing clues is the sweet potato. Sweet pota-toes are native to the Americas, cultivated in the Andes as early as 8000 BCE. After the Spanish reached Peru in 1532, sweet potatoes were introduced to Europe and brought to the Far East by Spanish merchants.* The funny thing is, even though sweet potatoes were introduced throughout the world by Europeans, Polynesians already

---

*Today, China produces more sweet potatoes than the rest of the world com-bined, and they're also popular throughout Africa and Southeast Asia. Despite their origin, the Americas grow less than 3 percent of all sweet potatoes.

81

had them when Europeans arrived—and had for a very long time. When Captain Cook landed in New Zealand, Maori farmers greeted him with arms full of sweet potatoes, and the British found row upon row of well-managed sweet potato fields. The New Zealanders had been sweet potato farmers for centuries.* How could this be?

Sweet potatoes seem to have moved across the Pacific in the opposite direction of Polynesian settlers. By around the year 800, Polynesians reached their easternmost outpost at Easter Island. This is precisely the time that we see sweet potatoes appearing in the archaeological records of Polynesian societies. By the year 1000, sweet potatoes appear in the Cook Islands, about halfway across the Pacific. One explanation is that somehow sweet potatoes floated across the Pacific to a Polynesian island, where they were picked up and traded throughout the rest of the Pacific. But this is almost certainly wrong. Sweet potatoes rot very quickly in water and won't germinate after they get waterlogged. It's extremely unlikely that sweet potatoes could have floated across thousands of miles of ocean. They're too big to be carried by birds. The obvious conclusion is that sweet potatoes were somehow carried from the Americas into the Pacific by humans in boats, long before Europeans reached the Americas.

Coconuts are a similar case. They float and are able to survive short journeys between tropical islands on their own but rot quickly at sea. Prior to European arrival, coconuts were already widespread from the Pacific to India, where they probably originated. Vasco da Gama brought coconuts to Europe from India in the early 1500s, but they didn't yet exist along the west coast of Africa. In fact, the uneven but widely spanning distribution of coconuts perfectly corresponds to the known movements of human seafarers, with one notable exception: coconuts were already established on the Pacific coast of the Americas before Europeans arrived.†

---

* Genetic drift analyses confirm that the sweet potatoes cultivated in Polynesia were introduced separately (and earlier) from the ones that Europeans picked up in the Americas.

† To further this point, coconuts failed to cross even smaller bodies of water where humans had not traveled—from Indonesia to Australia, for example, where they were only later introduced by humans.

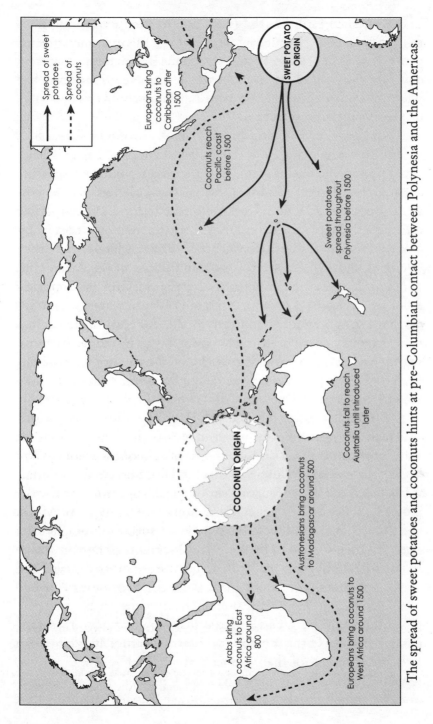

The spread of sweet potatoes and coconuts hints at pre-Columbian contact between Polynesia and the Americas.

83

This is the same story as the sweet potato, except coconuts seem to have moved in the opposite direction. This doesn't conclusively prove that Polynesians visited the Americas—a few specimens could possibly have survived an ocean journey of thousands of miles on their own—but the distribution and timing of coconut dispersal is awfully suspicious.

Though we have no conclusive proof that Polynesians landed in the Americas before Columbus, we do know that they were certainly capable of this feat. The Americas are only a third more distant from Easter Island than the nearest Polynesian settlement in the opposite direction. Compared to Pacific islands, pinpricks on a vast ocean, continents running from north to south across the globe would hardly be inconspicuous. When we consider the difficulty of locating remote specks of land across thousands of miles of ocean, can we even doubt that Polynesians found the Americas? All it would have taken is for a single Polynesian navigator to sail eastward in a straight line and then be able to return home. In fact, it seems likely that Polynesians would have visited the Americas multiple times, even on an ongoing basis. Would they have left any evidence? Perhaps not much. Even if a few individuals had stayed to settle in the Americas, any hint would likely have been subsumed into the melting pot of much larger American civilizations.

The other possibility is that Native Americans went the other way, into the Pacific. This theory was proposed in the mid-1900s by the Norwegian explorer-adventurer Thor Heyerdahl, although it has fallen out of favor based on linguistic, genetic, and archaeological evidence. We generally assume that Native Americans had little, if any, naval technology beyond simple canoes, but this doesn't actually seem to be the case. Our image of Native Americans as landlubbers actually arose later. In his foundational survey of American societies of 1815, the German botanist Adelbert von Chamisso declared that "no American society was ever a people of sailors." This seems a pretty sweeping and definitive statement, particularly for someone who arrived three hundred years after the conquest of the Americas. While Native American vessels of Chamisso's day may not have left much to the imagination, we shouldn't take that to mean that Native Americans were always incapable of sea travel. After all, when Europeans arrived at Easter Island, its inhabitants were incapable of sea travel, even though their ancestors had crossed thousands of miles of ocean to get there.

Part of this prejudice probably stems from the types of ships natives built, such as seagoing rafts. Actually, "raft" may bring the wrong image to mind. Think flat-bottomed sailboats, up to ninety feet long and thirty feet wide, with a cargo capacity of several full-grown elephants, and you'll have a pretty good idea of the kind of "raft" we're talking about. In modern times, adventurers have crossed both the Atlantic and the Pacific using raft-based designs. Some South American rafts had shelters and cooking fires on board and remained at sea for weeks. In fact, the Spanish encountered Incas at sea before they met them on land. When a reconnaissance mission sailed down the Pacific coast from Panama in 1526, it was surprised to find a large craft approaching, in a region where there should have been no Europeans.* The Spanish intercepted the ship and took several of its crew hostage. They described it as a large seagoing raft as big as their own vessel, made of logs lashed together, covered by a wooden deck, with sturdy masts supporting broad sails.

These types of ocean rafts were common along the Pacific coast of South America. Even the Galápagos Islands, hundreds of miles out at sea, may have been within the Incan sphere. The Incas had no writing, but a Spaniard named Pedro Sarmiento de Gamboa assembled a comprehensive history of the Incas in 1572 by consulting locals who'd been alive before Europeans' arrival. One of the accounts describes a sea voyage undertaken by the Incan emperor Tupac Yupanqui,† who had ruled a century earlier. According to the story, strange foreigners arrived from the west with tales of islands out at sea filled with treasures and gold. Not satisfied with his terrestrial domains, Tupac Yupanqui ordered the construction of a fleet of rafts and embarked with twenty thousand chosen sail-

---

* The Spanish ship was built on the Pacific coast of Panama, which had been reached only thirteen years earlier.

† The rapper Tupac was named after the last Incan emperor, Tupac Amaru, who was executed by the Spanish after leading a rebellion against their rule.

ors to head out into the western ocean. There, according to legend, they visited "several islands" before returning to South America with many treasures.

Did the Incas sail to the Galápagos, or possibly somewhere beyond? There is no conclusive evidence, but based on what we know of their naval technology, this would have been possible. The Galápagos remained unsettled, but archaeological surveys have found ceramic fragments and a flute that might be of Incan origin. The foreigners in the story are also suggestive, since Polynesians were clearly able to reach South America from the west around that time. Indeed, with such a small separation between two cultures possessing seafaring technologies, it's almost impossible to believe that Polynesians and South Americans didn't interact at some point, if only to a limited degree.

Thor Heyerdahl went further, proposing that Polynesians originated not in Asia but in the Americas. After fighting the Nazis in the Norwegian resistance during the Second World War, Heyerdahl tapped into postwar energies to prove that it was possible for ancient people to cross vast oceans. In 1947, Thor and five companions traveled to Peru and built a seagoing raft based on descriptions of Inca designs. Then, using only technology available to the Incas, they successfully sailed their raft, *Kon-Tiki*, five thousand miles for 101 days from South America to Polynesia. To their surprise, the raft proved to be highly maneuverable, and fish congregated between the balsa logs in such numbers that they had no trouble catching them for food and water. The expedition became a global sensation, spawning numerous award-winning books and films. It also may have contributed to the postwar craze for everything Pacific, including mai tais, cocktail umbrellas, and tiki torches.

Although Heyerdahl's theory of Polynesian westward settlement has been disproven by subsequent evidence, he did succeed in demonstrating that pre-European contact was not only possible but actually practical. Twenty years later, he went on to re-create the same feat in the Atlantic. Based on drawings from ancient Egypt, he built a boat from papyrus reeds with the help of traditional builders from Lake Chad in Central Africa. Named *Ra* after the Egyptian Sun god, this was a broad seagoing ship, bound tightly together for strength and rigidity. Thor launched it into the Atlantic on the coast of Morocco and set off for the Americas with a crew of seven. After a few weeks at sea, *Ra* started to take on water. High seas and storms eventually spelled doom for *Ra*, but not before it had sailed a distance of more than four thousand miles. Forced to abandon ship, the crew was saved by a passing yacht a few hundred miles short of the Caribbean. Belatedly, they discovered that they'd neglected a key design feature of Egyptian ships: a supporting cord that kept the front of the boat high up out of the water.

After correcting this flaw, a new ship, dubbed *Ra II*, set sail from Morocco the next year. This time the voyage was a complete success, making landfall in Barbados on the opposite side of the Atlantic. Heyerdahl had once again proved that ancient people were capable of crossing the most formidable oceans in the world using simple technology. Although there's no evidence that the Egyptians were interested in crossing the Atlantic, the voyages of *Kon-Tiki* and *Ra* illustrate what would have been possible. If the ancient Egyptians and Incas were capable of crossing oceans, we can be confident that more experienced seafarers such as the Polynesians, Phoenicians, and Greeks could have done the same.

There may have been many unrecorded early contacts across both the Pacific and the Atlantic, but unless and until we find firm archaeological or other evidence, we can't say conclusively when, where, or even whether such contacts might have occurred. Nevertheless, there are plenty of clues. For example, the Chumash people of California and Mapuche people of South America both developed sophisticated sewn-plank canoes unlike any others in the Americas but very similar to Polynesian models around the time when transpacific contact would have been possible. Tools, axes, earthen ovens, and fishing traps similar to those used by Polynesians have also been found in the Americas, particularly at Chiloé Island on the Chilean coast, which is populated by the Mapuche. Can it be pure coincidence that the most Polynesian-like technologies in the Americas are found opposite the nearest Polynesian island (Easter), a mere two thousand miles away?

There are also linguistic clues hinting at cultural interchange. The Chiloé Islander word for sewn-plank canoe, *"kia lu,"* may come from the Polynesian "*tia lao*," for "sew" and "long," or the Hawaiian "*kialoa*," for a long, light canoe. Similarly, the Chiloé word for "ax," "*totoki*," is almost identical to the Easter Islander word for ax, "*toki*." These linguistic connections are especially intriguing because Chiloé Island is one of the easiest parts of the Americas to reach from Polynesia, a straight shot in the reliable winds of the Roaring Forties. Indeed, it's plausible that Chiloé Islanders are the product of significant mixing of Polynesian and South American peoples, at least culturally, if not genetically. A linguistic connection is also suggested by the sweet potato. As we've seen, sweet potatoes originated in the Americas but were scattered across Polynesia prior to European contact. Among Polynesian cultures, the word for "sweet potato" is based on the proto-Polynesian "*kumala*," like the Hawaiian "*uala*" or Maori and Easter Islander "*kumara*"—virtually identical to the Quechuan* "*k'umar*." It seems that if sweet potatoes floated across the ocean, they somehow came with a soundtrack.

There are also genetic clues. As they traveled through the Pacific, Polynesians brought along chickens for meat and eggs. Before the arrival of Europeans, there shouldn't have been any chickens in the Americas, but a 2007 study radiocarbon-dated chicken bones found in Chile to between 1304 and 1424, well before Europeans' arrival. Particularly since the bones were from the Pacific coast, this hints at the possibility of early contact with Polynesia. Indeed, the chicken DNA sequences proved exact matches to chickens from the same period found in Samoa, Hawaii, and Easter Island. However, radiocarbon dating isn't perfect, so the jury is still out on the chicken connection. On the human side, several studies have found supposed traces of Polynesian genes in the Americas, or South American genes in Polynesia, but all of these studies are disputed and none are entirely conclusive.

Part of the problem with confirming early contacts is that we don't have a very clear picture of pre-Columbian America. Few Native Americans used writing, and most were wiped out by European contact. What we do know generally comes from sparse

---

* Quechua was the language of the Incas and is still spoken in South America today.

archaeological evidence or severely biased European accounts. Early Spanish accounts were often written to justify their conquests, enslavements, and forced religious conversions. One story that falls into this category may be the "white god" myth. Cortés and his men supposedly reported being received by the Aztecs as the god Quetzalcoatl, who was predicted to return in the exact year they landed in Mexico. In Peru, Pizarro and his men similarly reported being received by the Incas as the return of their creator god Viracocha. The similarity is striking: Quetzalcoatl and Viracocha were both interpreted (at least by the Spanish) as tall, bearded, white-skinned, blue-eyed gods who were expected to return from across the great ocean.

Some versions of these accounts are conspicuously detailed. The original Quetzalcoatl had supposedly come from across the sea clad in a coarse black garment. Seen as a moral teacher, the god enacted new laws and tried to ban human sacrifice. He also introduced variants of baptism and confession, which the Spanish claimed were already widespread in the Americas before the arrival of Catholicism. Followers of the rival god Huitzilopochtli eventually rose up against the foreign Quetzalcoatl, who fled across the eastern ocean into the land of dreams—but not before prophesizing that his angry white brothers would one day return to conquer Mexico. The return of Quetzalcoatl was anticipated on the Aztec day of 1-Reed at the conjunction of their 365- and 260-day calendars (which occurred every fifty-two years). As it so happened, Cortés landed in 1519, precisely on that date.

The extreme coincidences strongly suggest that these stories were either partly or entirely fabricated by the Spanish. The details are just too convenient. But it's possible that at least a few elements were loosely based on indirect transatlantic contact, perhaps in the form of a semi-mythical legend spreading across the Americas triggered by Viking contact. Interestingly, most Spanish accounts describe the foreign gods as having blond hair and blue eyes, which would be unusual for the Spanish but downright characteristic of the Vikings. The description of Quetzalcoatl's cloak better fits what would be expected of a medieval Viking's garment than a contemporary Spaniard's.

Or could it have been someone other than the Vikings? We've seen that plenty of Old World people, from the Egyptians to the Irish, possessed the capability of crossing oceans. While this doesn't confirm that they *did*, given thousands of years of capability, it's hard to say that it never happened, even if accidentally. Leaving from Southern Europe, a ship caught in prevailing currents practically can't fail to reach the Americas in a few months on a dead drift. Add prevailing winds and you cut the crossing time to a matter of weeks (it took Columbus twenty-nine days). It's only 1,600 miles across the Atlantic from Cape Verde—far shorter than distances regularly traveled by Polynesians in the Pacific. Once you start sailing west, it's hard to miss the line of continents spanning the entire globe from north to south.

If a sailing ship lost its mast in a storm, might it not have been carried across the ocean, even if its crew weren't sufficiently stocked with provisions to survive the crossing? Columbus's accounts show that he was inspired by carved driftwood and other flotsam of human origin washing up on European beaches "from lands across the sea." Would it not have been possible for entire vessels, either derelict or with crews clinging to life, to be carried in the other direction? Once landed, a crew might not have been able to make repairs and so might have faded into the great American melting pot, while igniting some fascinating stories.

Did early contacts take place between the Old and New Worlds? We can't say for sure. Contact between Polynesia and South America seems very likely, but transatlantic contact, other than to a limited degree through the Vikings, is more uncertain. The best we can say is that although no continuous and sustained contact is evident, accidental and sporadic contact was certainly possible, and perhaps even probable. It's a question that may forever live in the realm of conjecture. As we've seen throughout history, people were moving around a lot more than we're used to thinking, without always leaving traces of their travels. Lacking firm supporting evidence, we must remain skeptical about the details but open-minded about the possibilities.

8 | THE OTHER MEDITERRANEAN

We now turn to the other sea at the heart of the world.* This one is more vast, more ancient, and in some ways more important than the sea lapping at the shores of Greece and Rome. From ancient times, the Indian Ocean acted as a conduit between far wealthier civilizations than those of the Mediterranean, carrying luxury goods, technologies, and ideas between Africa, the Middle East, India, China, and Southeast Asia. Although Europe eventually rose to explore and conquer much of the world, its preeminence was far from clear a mere five centuries ago. Typical Western accounts of exploration introduce the Indian Ocean with the arrival of Vasco da Gama,† but the people of this region were master seafarers for thousands of years by the time they were "discovered" by the Portuguese.

Indeed, the Indian Ocean was in many ways the true center of the world, a center from which Europe was excluded during the Middle Ages. It is this very exclusion that prompted the Age of Exploration. The voyages of Vasco da Gama and Columbus would have been meaningless if undertaken by an Indian navigator. There was no need for Indians to make such a voyage, for they were already living in the place Europeans were trying to reach. At the

---

* In Latin, "Mediterranean" = "sea at the center of Earth."
† The Portuguese navigator who first reached India by sailing around Africa in 1497.

time, Asia was so wealthy relative to backward Europe that European powers constantly struggled to find trade goods that Asians wanted to buy. Europeans often took goods by force, not only because they could, but also because they had nothing of value to offer. This was the root of the 1839 Opium War between the British Empire and China. Britain didn't have anything China wanted to buy, so they forced their way into Chinese markets by demanding to trade cheap opium from India for valuable Chinese silks and porcelain. When he arrived in India, Vasco da Gama brought worthless European trinkets to trade for spices. Upon the explorer's return, King Manuel I of Portugal, reflecting on the relative poverty of Europe, remarked, "It is not we who have discovered them, but they who have discovered us."

Indians, Arabs, Africans, and other people of the region have been sailing the Indian Ocean since the beginning of civilization. The ancient Sumerians used ships not unlike those of ancient Egypt to transport grain, pottery, copper, tin, dates, and pearls along the Tigris and Euphrates and into the Persian Gulf. Soon vessels of Middle Eastern and Indian design were engaged in a three-way trade between Mesopotamia, the Harappa civilization of India (2600–1900 BCE), and the Old Kingdom of Egypt, which built the Great Pyramid. The Indian Ocean's relatively calm waters made it easier to sail than the Atlantic or Pacific, and reliable monsoon winds facilitated trade, with ships sailing east during the summer and west in the winter.

Indian Ocean trade relied on dhows, wooden cargo vessels that are still used across a swath of the world from Africa to Indonesia. Dhows originated in the region over two thousand years ago, introducing a number of innovations that were later copied by European sailing ships. With raised hulls and sharp pointed bows, dhows boast superior handling, even in rough seas, and their triangular sails make it possible to sail into the wind. Early dhows used sails made of woven palm fronds, but soon dhows were propelled by sails of high-quality Indian or Egyptian cotton. Despite their innovative design, dhows could be simply constructed using planks bound together by palm fibers. Europeans who encountered them were amazed that such seaworthy vessels could be held together without any nails or rivets, but in fact, exposure to the sea causes the wooden planks to swell, binding them ever tighter.

The maritime history of the Indian Ocean was largely suppressed by Europeans who sought to dominate the ocean's trade routes, but it's possible to reconstruct much of it. When the Portuguese arrived, prosperous city-states flourished along Africa's east coast, boasting excellent natural harbors at which to load exotic goods. Merchants from India, Arabia, and as far away as China came to trade, and many chose to stay, marrying local women. Gradually, the "Swahili" culture developed, with a language (the most widely spoken in East Africa) that contains a significant fraction of Arabic- and Hindi-derived words (in fact, "Swahili" comes from the Arabic for "coastline"). Sofala, Mombasa, Malindi, and other African trading posts grew into major cities. The famed traveler Ibn Battuta visited the East African city of Kilwa (Tanzania) in 1331 and described its immense buildings and countless splendors. Although these cities were known to the Romans and Greeks,* their very existence came as a surprise to the Portuguese when Vasco da Gama stumbled upon them in 1497.

Trade networks into the African interior brought gold, ivory, minerals, and other products to East African ports from inland cities such as Great Zimbabwe. A major city of tens of thousands located hundreds of miles from the coast, Great Zimbabwe boasted a sprawling royal palace surrounded by walls twenty feet tall. The city was recorded by early Portuguese explorers, but the fact that such a metropolis could be built by Africans was denied for decades by the apartheid government of Rhodesia. The city has since been adopted as a national monument of the modern nation of Zimbabwe, which was also named in its honor. (The "Great" distinguishes the capital from hundreds of towns across the region, collectively known as "zimbabwes" and together forming the Kingdom of Zimbabwe.)

To the east, sailors from India had been perfecting their navi-

---

* The two-thousand-year-old Greek travel guide to the Indian Ocean, *Periplus of the Erythraean Sea*, describes East African ports as far south as Mozambique, south of the equator on the continent, opposite Madagascar.

gational skills since ancient times. Chandragupta Maurya (reigned c. 322–298 BCE), who united the subcontinent in the wake of the power vacuum left by Alexander the Great, founded the Indian navy and established overseas trade relations. His grandson, the emperor Ashoka (reigned c. 273–232 BCE), established settlements in the Maldive Islands and sent diplomatic embassies to Europe and China. Indian sailors were early adopters of celestial navigation and pioneered the development of detailed navigational charts. Regular visitors to Egypt, Indian travelers left Tamil and Sanskrit inscriptions in the Temple of Luxor. At least four diplomatic missions from India to the Roman Empire are recorded by Pliny the Elder, and Indian merchants were not uncommon in Rome itself.*

Maritime expansion brought not only trade but ideas. Buddhism originated in India but was spread to Southeast Asia by Indian and Sri Lankan sailors. This is why Indonesia, Malaysia, and Thailand share a cultural connection to Sri Lanka and India, including languages and architecture, even though they aren't connected overland. Trade also introduced new forms of cuisine to neighboring cultures, which is why Indian spices and curries are essential elements of Indonesian, Malay, and Thai cooking. As the commercial hub of the world, India grew wealthy exchanging goods from across the Old World as far as Europe to the west and Japan and China to the east. So much trade flowed through India that it often became impossible for foreigners to discern the origin of goods. Europeans wanted to get to "India" to buy spices. What they failed to realize was that most spices weren't grown in India but merely traded there.

Although some spices, like pepper, came from India, most were brought there by seafarers from farther east. They traveled the so-called Cinnamon Route, hauling not only cinnamon but also

* One sign of the Indo-Roman connection is the fact that both had identical meanings for days of the week: Monday (Moon Day) = "Somwar" ("Soma" = "Moon" in Hindi); Tuesday (Mars Day) = "Mangalwar" (Mangal = Mars in Hinduism); Wednesday (Mercury Day) = "Budhwar" (Budha = Mercury in Hinduism); Thursday (Jupiter Day) = "Guruwar" (Guru = Jupiter in Hinduism); Friday (Venus Day) = "Shukrawar" (Shukra = Venus in Hinduism); Saturday (Saturn Day) = "Shaniwar" (Shani = Saturn in Hinduism); Sunday (Sun Day) = "Raviwar" ("Ravi" = "Sun" in Hindi).

spices such as cloves, nutmeg, and mace from their only sources in the Indonesian archipelago—the first link in a ten-thousand-mile chain that brought spices to the dinner tables of wealthy Europeans from Roman times until the Middle Ages. The people who followed this route were the Southeast Asian seafarers whom we met earlier sailing across the Indian Ocean as far as Madagascar, relying on navigational techniques similar to those of their Polynesian cousins in the Pacific. By around the year 400, contact with India had introduced Southeast Asia to Hinduism, Buddhism, and Sanskrit writing,* which blended into local traditions and languages.

These seafarers put their skills to use fueling the spice trade. Until a few centuries ago, mace, cloves, and nutmeg were only produced in a small clutch of islands in eastern Indonesia (the Bandas and Moluccas). Their export was tightly controlled by local chiefs to drive up prices. Eventually, maritime empires emerged to control the trade routes, the most important being Srivijaya and Majapahit, politically autonomous thalassocracies based on the sea (like Phoenicia or ancient Greece). While they didn't necessarily produce the spices, Srivijaya and Majapahit did control the first link in the spice route until the arrival of Arabs and Europeans centuries later. These foreign conquerors suppressed much of the history of this region, so that even the existence of these maritime empires was largely unknown until recently.

Srivijaya was a mainly Buddhist empire that rose in the seventh century, when increasing appetites for spices brought wealth to Indonesia. This ancient splendor can still be felt at the magnificent Borobudur Temple on Java,† the largest Buddhist complex in the world, featuring 2,672 carved panels depicting Buddhist cosmology (i.e., the birth, life, and death of the universe). In time, Srivijaya declined and was eventually replaced by Majapahit, a

* Sanskrit is like the Latin of India and Southeast Asia. It's still widely used for religious and ceremonial purposes but rarely spoken. Nevertheless, it's considered one of twenty-two official languages of India and historically played an important role as a widely understood common language across the region.

† The most populous island in the world and main (but not largest) island in Indonesia—and also the origin of the colloquial name for coffee.

Buddhist-Hindu empire with a peculiar origin. In 1292, Kublai Khan, the Mongol ruler of China, sent an emissary "inviting" Java to pay tribute. The Javanese refused—never a good idea when a khan asks nicely. In response, Kublai ordered a thousand-ship invasion of Java, the largest long-range Chinese seaborne invasion in history.* Majapahit first sided with the Mongol invaders to help them eliminate their rivals and then turned against them with a surprise attack to seize control of Java.

Majapahit ended up controlling most of the region's trade. The largest empire in the history of Southeast Asia, it defined the boundaries of modern Indonesia. The fourth-most populous country in the world today, Indonesia spans three thousand miles and encompasses twenty thousand islands. Around 1324, the Italian friar Odoric of Pordenone visited the Majapahit court on Java, where he received a warm welcome. He would later describe the powerful kingdom and its extensive maritime network.† Majapahit maintained diplomatic relations with China and was visited on several occasions by the armadas of Zheng He between 1405 and 1433. Around this time, Islam, which had been introduced in Majapahit by Arab merchants attracted to the lucrative spice trade, spread throughout the empire. Soon, new Muslim rulers of Singapura (Singapore), Malaysia, and Sumatra overran most Majapahit territory. The remaining vestiges were absorbed by the Portuguese and Dutch a century later.

This brings us to the rise of the Arab merchants, who held a virtual monopoly over Indian Ocean trade by the time Europeans arrived in the early 1500s. As we've seen, Middle Easterners were

---

* Only the second Mongol invasion of Japan may have been larger, but that was over a much shorter distance. Today, there is tension in the South China Sea over territorial disputes involving China. The Mongol invasion of Java represents a projection of Chinese power approximately three times as far—around the distance from Ireland to Newfoundland. Meanwhile, Mongol armies based in Poland were invading Central Europe on the other side of the world.

† Odoric apparently sailed on a junk dispatched by the Chinese imperial court, where he was an honored guest, following in the footsteps of Marco Polo. Along the way, he may also have landed in the Philippines 197 years before Magellan, which would make him the first European to visit these parts of Southeast Asia—although it's likely that a Greek or Roman got there first.

involved in this trade from the beginning, but after the rise of Islam in 632 they pursued commercial activities with vigor. Muhammad was himself a traveling merchant, following caravan routes through Arabia. As a result, trade has always featured prominently in Islam. Although the Muslim world was rarely unified, adventurous merchants enjoyed an unusual degree of mobility throughout the region of Muslim rule, stretching from Morocco to the borders of China.

This traveling tradition was encouraged by the requirement that every Muslim make a pilgrimage to Mecca during their lifetime. For the medieval world, this was an unprecedented gathering of people who wouldn't have met in any other way, connecting pilgrims from as far away as Southeast Asia and Spain. In this way, every year millions of Muslims departed their homes on long journeys. But perhaps no journey was as epic as that of Ibn Battuta, who set off on pilgrimage to Mecca from Morocco in 1325 at the age of twenty-one and just kept going for the next twenty-five years. None of this was planned in advance, so we may suppose his family assumed him dead until he suddenly reappeared a quarter century later. Ibn Battuta may indeed be the human who has walked the farthest—over seventy-five thousand miles, the equivalent of three times around the world.

What motivated Ibn Battuta's travels? He explained it as a restless desire to understand the world. His journey became a sort of spiritual quest, with a self-imposed rule that he could never retrace his steps if he could see more by going a different way. Ibn Battuta wasn't a merchant but, rather, a *qadi* (Islamic judge) who made a living by practicing his craft along the way, while also benefiting from a strong Muslim tradition of hospitality to guests. As he went, people would ask him to perform errands. In Mecca he met a man who asked him to convey greetings to his brothers in India and China, and incredibly, he was eventually able to fulfill this request across thousands of miles. Later, rulers would solicit advice based on his vast geographical knowledge. They wanted to know what life was like in the next kingdom over. What were the latest ideas of governance, technologies, fashions, and trends?

Ibn Battuta traveled throughout the Middle East, Russia, Afghanistan, and India, and eventually all the way to China. Other

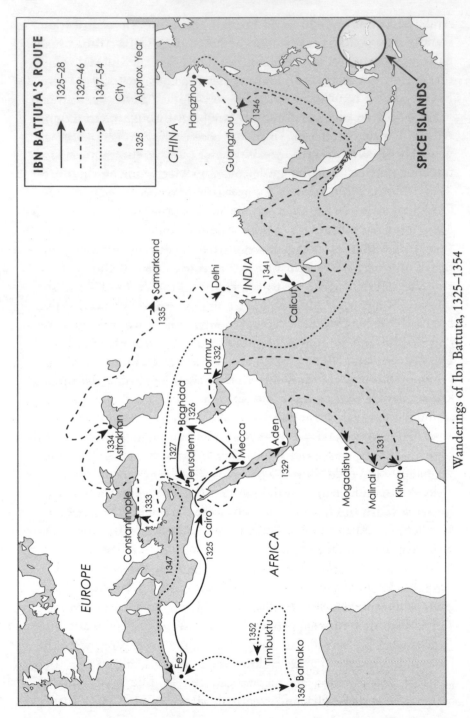

Wanderings of Ibn Battuta, 1325–1354

IBN BATTUTA'S ROUTE

1325–28
1329–46
1347–54
● City
1325 Approx. Year

CHINA
Hangzhou
Guangzhou 1346

SPICE ISLANDS

Samarkand 1335
Delhi
INDIA
Calicut 1341

Hormuz 1332
Baghdad 1326
1327
Astrakhan 1334
Jerusalem
Mecca
Aden 1329
Constantinople 1333
Cairo 1325

EUROPE
1347
AFRICA

Mogadishu
Malindi 1331
Kilwa

Fez
Timbuktu 1352
Bamako 1350

Arabs had been to China, and indeed Arab sailors were living in some of its southern ports by this time. Ibn Battuta described China's magnificent cities, canals, ships, and technologies but recorded that he felt out of place in a distant land for the first time there. Eventually, he returned home to Morocco but stayed only a few years before setting out on another adventure, this time across Africa. The highlight was Timbuktu, capital of the Mali Empire, which became fabulously wealthy as a caravan stop for goods traveling between Europe and Africa. In fact, its king at the time, Mansa Musa, is regarded as the wealthiest human in history—at least three times richer than Jeff Bezos, inflation adjusted. When Musa went on pilgrimage to Mecca (in 1325, precisely the same year as Ibn Battuta), his procession reportedly included twelve thousand slaves clad in fine silks and a train of camels hauling enough gold to depreciate currencies along his path for a decade.

After stopping in Mali, Ibn Battuta followed the course of the Niger River (something no European would manage for more than four hundred years) and observed that the river contained hippopotamuses. These had been described in ancient times ("hippopotamus" = "river horse" in Greek) but weren't widely known beyond Africa. Like later European explorers, Ibn Battuta assumed the Niger was either a tributary or in fact the same river as the Nile, since it followed an easterly course toward Egypt and both rivers contain hippos—but in fact, the Niger makes a sharp turn south and comes no closer than two thousand miles to the Nile. After wandering through the wealthy cities of West Africa for a few years, Ibn Battuta received a command from the sultan of Morocco to return home and dictate an account of his travels for posterity. The resulting volume of travel writings, *A Gift to Those Who Contemplate the Wonders of Cities and the Marvels of Traveling*, has inspired numerous books, films, poems, and even video games across the Muslim world, where "Ibn Battuta" has become a byword for a frequent traveler.*

* He is also known as the namesake of the world's largest themed shopping mall, the Ibn Battuta Mall in Dubai, which features sections named after regions he traveled through: Egypt, India, China, Andalusia, Persia, and Tunisia.

Not merely a traveling tradition but a seafaring one is ingrained in Islam, as exemplified by "Sinbad the Sailor," a classic maritime adventure tale in the vein of the *Odyssey*: it features treacherous storms, exotic lands, magical creatures, and giant whales masquerading as islands. The story was first introduced in *One Thousand and One Nights*, the narrative of a mad sultan who marries a new wife every day and then executes her the next morning. On their wedding night, a woman named Scheherazade begins telling a fascinating tale but doesn't finish it. The sultan, curious to learn how the story ends, postpones her execution to hear more. The next night, as soon as she finishes the story, Scheherazade begins a new tale, and the sultan is forced to spare her once again. Thus it goes for 1,001 nights, and 1,001 tales (the original serial cliff-hanger).*

Soon after the advent of Islam, Arab sailors started to venture beyond the Middle East and into the trade routes of the Indian Ocean. Within a few centuries, seafaring Arabian merchants were a common sight as far as China, where mosques dotted coastal seaports long before the arrival of Europeans. Arabs mainly sailed the simple and rugged dhows common in the Indian Ocean for two millennia. They built forts in Indian harbors, where they established merchant guilds that negotiated favorable trade deals with local potentates. Arabic became so prominent in the Indian Ocean that Vasco da Gama brought along Jewish interpreters who'd learned to speak Arabic. In fact, when Columbus reached the New World, he took it as a very bad sign that the natives couldn't understand Arabic, because this implied that he wasn't near the trade routes of Asia.

There was an overland route between Asia and Europe along the Silk Road, but the sea route through the Indian Ocean and up the Red Sea to Egypt was much more efficient. Once in Egypt, exotic Asian goods traveled a few days by caravan to Alexandria, where they were loaded by Venetian sailors for distribution in Europe. This trade made Venice wealthier than any other state in Europe, but it was merely the last link in a network that spanned

---

* Other stories from *One Thousand and One Nights* you may recognize: "Ali Baba and the Forty Thieves" and "Aladdin."

almost half the globe. Meanwhile, Europeans had only the vaguest notion of where the silks and spices they enjoyed came from. It seems rather odd to us today that the kitchen spices we can buy at any supermarket for a few dollars once ruled the world's economy, causing the rise and fall of nations. Why were spices so important? It wasn't actually to hide the flavor of rotting meat (as has often been supposed). It was mostly a matter of prestige, a reflection of drab medieval life in obscurity at the forgotten edge of the world. European nobles who could afford to serve exotic meals could temporarily rekindle memories of better times, when Greeks and Romans lived in luxury, trampled Asian empires, and stood triumphant on the world stage.

Medieval European cooking was, by our standards, bizarre to say the least. Recipes would mix whatever spices were available to create flavor pairings that would seem rather incongruous to the modern palate. Spices ranged from the types we know today to the exotic "mummy spice," made of ground-up Egyptian mummies and supposed to convey healing powers. Savory and sweet weren't segregated into main course and dessert but just served whenever. Pepper had been popular since Roman times—so much so that the barbarian Alaric the Goth demanded three thousand pounds of peppercorns in exchange for not sacking Rome. But the pairing of pepper with salt on dinner tables traces to the arbitrary culinary preferences of Louis XIV, who rejected all seasoning beyond a pinch of salt and pepper, and banished most spices from his kitchen. As nobles across the continent followed suit, it set the pattern for European dining we know today.

We sometimes envision Europeans of the Age of Exploration entering a backward region devoid of maritime trade. In fact, it was the existence of an established network that drew Europeans, who entered a crowded region armed with bigger guns and the intent to conquer. In the Indian Ocean, this wasn't an easy task. Vasco da Gama received a chilly reception in East Africa in 1497, and the first navigator he hired tried to wreck his fleet on a shallow reef. Upon reaching India, da Gama met Arab merchants, a few of whom angrily demanded (in Spanish) to know why the Portuguese had come. Soon, the Arab merchants of India organized local resistance against the Portuguese, starting with an attack on da Gama's second expedition in 1502 in the Battle of Calicut. The conflict became a prolonged struggle, as the Portuguese viciously competed with established interests for control of the Indian Ocean. In this struggle for trade at the center of the world, European victory was, at least initially, far from a foregone conclusion.

The most serious challenge came from the Ottoman Empire. Larger, wealthier, and arguably more advanced than any European power of the time, the Ottomans rose from scattered Turkic tribes and in 1453 conquered Constantinople, the last remnant of ancient Rome and the capital of the Byzantine Empire. This event actually sparked the Age of Exploration, since it caused Europeans to worry that they might be cut off from Asian trade by Ottoman control of the Silk Road. The Ottomans also posed a direct existential threat, carving out a massive swath of territory from Greece to Hungary and menacing all of Christendom with their sieges of Vienna in 1529 and 1683.* Not merely a land power, the Ottoman Empire also operated the largest navy in the world from their territories that stretched two-thirds of the way around the Mediterranean from Greece to Morocco. If the Ottomans could have somehow transported their magnificent fleet overland to their ports in the Red Sea or Persian Gulf, they could easily have eliminated the tiny Indian Ocean fleets of the upstart Portuguese.

Ottoman naval power during the first half of the 1500s was virtually unchallenged in the Mediterranean, and it remained a major force long after the Ottomans' defeat at Lepanto in 1571 at the hands of a Christian coalition. At times, Ottoman power stretched far out into the Atlantic. There were raids on the Canary Islands in 1585, on Madeira in 1617, and then on coastal towns in southern England in 1625. In 1627, Ottoman pirates established a base at Lundy Island (a mere twelve miles off the coast of Britain!), plundering ships and coastal towns from England and Ireland to as far afield as Scandinavia. During this time, Ottoman vessels ventured

---

* It's been suggested that coffee was introduced to Europe during the 1683 siege, when Europeans looted the baggage of the grand vizier, who brought some along. This story may be apocryphal, but the first coffeehouse in Vienna did coincidentally open a few years later.

all the way to North America, where they were sighted off the coast of English colonies in Newfoundland and Virginia.*

The Ottomans are sometimes criticized for not expanding overseas or even colonizing the Americas. This entirely misses the point. The only reason the Spanish sailed to the Americas, or the Portuguese sailed around Africa, was to break into the part of the world already occupied by the Ottomans. From a European perspective, Africa and the Americas were not ends in themselves but merely obstructions. The Ottomans were well aware of the discovery of the Americas. In fact, one of the largest and most accurate world maps was assembled by the Ottoman admiral† Piri Reis in 1513, depicting Europe, Asia, Africa, and the Atlantic coast of the Americas (only the Atlantic section has survived). But, perfectly established as they were to control the lucrative trade of Asia, why would Ottoman admirals choose to sail into the wilderness of the Americas? They were already precisely where they wanted to be, at the center of the world.

The Ottomans were so powerful that they were able to challenge all the Christian navies of Europe while simultaneously battling Portugal for control of the Indian Ocean. They never managed to transport their Mediterranean fleet overland across Egypt, but they did build fresh fleets of over a hundred ships in the Red Sea and Persian Gulf. Both these regions were chronically short on lumber, so it was no small feat to transport vast stockpiles of wooden planks from European forests, first by sea across the

---

* This was not the only time Americans came into contact with Ottomans. In fact, the original US navy was built for the express purpose of fighting the Ottoman-sponsored Barbary corsairs of North Africa, who enslaved American sailors after a US refusal to pay ransoms or bribes. With the slogan "Millions for defense, but not one cent for tribute," Congress in 1801 ordered six new frigates and declared war. Several battles ensued, including the storming of a ship in Tripoli Harbor by US marines—hence the line in the Marine Corps hymn "to the shores of Tripoli." This war was also the first time a US flag was raised on foreign soil, when a party of marines led a force of five hundred mercenaries across the desert from Egypt to capture the city of Derna (near Benghazi). Thus, an invasion of Libya was America's first foreign war. The heroic father of the American navy, John Paul Jones, also fought the Ottoman navy when he signed on as a mercenary admiral of the Russian Black Sea Fleet (people really got around in those days!).

† An Arabic word that entered our vocabulary via the Moors of Spain.

Mediterranean and then by caravan across the desert to shipyards at Suez and Basra. To support this effort, the Ottomans revived the dream of building a canal through Egypt that would link the Mediterranean to the Red Sea, and actually started digging in 1565, but the project was soon abandoned.

Between 1538 and 1553, the Ottomans organized four major naval expeditions against the Portuguese in India. These were concerted attempts to link up with local Muslim rulers and merchant guilds to eject the Portuguese from their fortresses at Goa and Diu. They narrowly failed each time, usually for diplomatic rather than military reasons. After the last expedition, the Ottoman admiral Seydi Ali Reis traveled to the Mughal court, where he tried to persuade the sultan (the most powerful ruler in India and also a Muslim) to join the Ottoman cause, but to no avail.* Had this alliance succeeded, the Portuguese wouldn't have stood a chance in India, but they benefited from divisions in both India and the Muslim world.

However, despite their failure to expel the Portuguese from India, the Ottomans were quite successful at sea, winning battles against the Portuguese using both sailing ships and the same nimble galleys that had proven so effective in Mediterranean waters since ancient times. At the same time, the Portuguese attacked all manner of shipping in the Indian Ocean, disrupting trade patterns that had existed for millennia. The Ottomans, as guardians of the holy cities of Mecca and Medina, considered it a sacred duty to protect ships transporting religious pilgrims, and these were frequently attacked by the Portuguese, often in brutal ways. Thanks to the Ottoman challenge, the Portuguese never succeeded in their goal of controlling the sea lanes, so Ottoman and other Muslim ships continued to sail the Indian Ocean in large numbers until modern times.

Portugal's presence in not only India but Southeast Asia, too, was threatened by Ottoman efforts. In 1569, the Ottomans sent a fleet of twenty-two ships with soldiers, cannons, and technical

---

* He probably met the future emperor Akbar "the Great," who was twelve at the time (one of the most important Indian rulers in history).

experts to support an attack on the Portuguese fortress of Malacca (Malaysia) by the sultan of Aceh (Sumatra). Although the attack failed, Aceh posed a constant threat to Portuguese control of the region, thanks in large part to the cannons produced in foundries established by Ottoman craftsmen. This Ottoman assistance contributed greatly not only to bolstering local defenses against European powers but also to the spread of advanced technology throughout the region. More than three hundred years later, when Aceh was attacked by the Dutch in 1873, the sultan asked for support from the Ottomans. The Ottomans were preoccupied with a rebellion in Yemen, but the fact that a Southeast Asian nation four thousand miles away would seek support from them against a modern colonial power is testament to their influence in the region.

The struggle for control of the Indian Ocean was a sideshow for the Ottomans, but it was a national effort for the Portuguese. Nevertheless, Portugal largely failed in its goal of controlling the sea lanes. While Portuguese ships had to carry cargos of spice more than twelve thousand miles around Africa, the Ottomans continued to use the much more direct route up the Red Sea into Egypt. Venice still loaded spices in Egyptian ports and continued to import an equal quantity into Europe via this route, even at the height of Portuguese power.* The Portuguese themselves recognized the superiority of this shorter route and hoped their efforts in the Indian Ocean would eventually lead to a direct invasion of Egypt. By this means, the Portuguese didn't want to outcompete the Ottomans so much as replace them, and in this they failed utterly. Egypt remained the shortest and most efficient link between Europe and Asia. This fact would soon motivate other European powers, such as Napoléon's France and then the British, to seize direct control of Egypt, culminating in the opening of the Suez Canal in 1869 to permanently bind East to West.

---

* In a Renaissance marketing campaign, Venice claimed that the long voyage around Africa in damp Portuguese cargo holds spoiled the aroma of spices.

9 | CHINA'S AGE OF EXPLORATION

We often speak of the rise of China as a recent phenomenon, but in fact, for the majority of human history, China has been the leading power in the world. It's only in the last five centuries that China experienced a decline in economic and military power relative to Europe and America. In classical times, China saw itself as a counterbalance to Rome at the opposite end of the world, approximately equivalent in both population and economic output. As we've seen, several official and unofficial Roman visitors made it to China, and at least one Chinese envoy, Gan Ying, traveled to the outskirts of the Roman Empire, publishing a report for the Chinese imperial court. In contrast to Rome, which eventually collapsed, China is essentially an ancient civilization that has survived and evolved over the centuries, dynasty to dynasty, with its language, culture, and legacy more or less intact.

Although primarily a land power on the eastern edge of Asia, China has a long history with sophisticated naval technology. Large multidecked vessels have navigated China's rivers for at least three millennia, since long before equivalent designs were imagined in Europe. The Qin dynasty* initiated overseas expansion, with a seaborne invasion of Vietnam and expeditions as far as India and possibly East Africa. The subsequent Han dynasty

* The first dynasty to unite China (221–207 BCE), which gave the country its name.

(206 BCE–220 CE) built the first large seagoing vessels, called "junks."* These introduced several innovations not adopted in the West for over a thousand years, including watertight compartments for durability, centerboards for stability, and sternpost rudders for efficient steering. The most recognizable feature of junks is the sails, fixed by many horizontal beams to look like a sort of segmented fan. This makes the sails slightly heavier but also more rugged. Nevertheless, junks are surprisingly nimble, roughly equivalent to European sailing ships of a similar size.

With the development of sturdy seagoing vessels, Chinese sailors embarked on regular expeditions into the Indian Ocean (where they probably encountered Romans, who also visited India around this time). In the late 700s, the Chinese geographer Jia Dan recounted voyages to India, Arabia, and East Africa in his *Route Between Guangzhou and the Barbarian Sea*. By the Song dynasty (960–1279), Chinese ships regularly sailed as far as Madagascar, Egypt, and into the Persian Gulf to trade with the powerful Abbasid caliphate in Baghdad (where they may have met Vikings). Chinese accounts accurately describe bustling Middle Eastern bazaars, minaret-style lighthouses, and harbors filled with triangular-sailed dhows. Meanwhile, descriptions of Chinese junks and sailors start to appear in the accounts of Arabs, who were embarking on their own eastward voyages to China.

Then, all of a sudden, in 1206, everything changed. In that year, a new terror rose in the steppes of Central Asia to unleash devastation on an unsuspecting world. Within decades, the Mongols would carve out one of the largest empires in history, stretching from Korea to Poland. The Mongols learned to ride from an early age and practically lived on horseback. Expert archers, they used long-range composite bows that could be fired at a gallop, propelled by layers of fibrous animal tendon. Mongol warriors rarely fought up close, preferring to stand off and shower their enemies with arrows. These tactics made them virtually impossible to defeat. More powerful but slower European knights and

* Junks are still used, although the name "junk" isn't Chinese, but comes from a Portuguese interpretation of a Malaysian word for "ship."

Chinese cavalry couldn't catch the nimble Mongolian horsemen, who simply stayed out of range. Their tremendous success on the battlefield allowed them to dominate the center of the world from China to Europe, inextricably binding East to West.

The Mongols mainly ate meat and drank mare milk, differentiating themselves from settled people by scornfully calling them "grass eaters."* Adept at warfare on the open steppes of Asia, they soon developed specialized tactics against walled cities. Often they would simply starve a city into surrender, which they accelerated by terrorizing the countryside to encourage as many people as possible to flee within the walls and consume remaining food supplies. Sometimes they'd tunnel under the walls,† or break river dams to flood a city. Later, the Mongols became experts at building siege engines such as catapults and rams, adopting the necessary construction techniques from Chinese and Western technicians. For the first time in world history, Chinese military advisers were employed against Western cities, and vice versa. In one instance during the siege of Kiev, Russian defenders were shocked by the spectacle of "flaming dragons" (Chinese rockets) assaulting their fortifications.

The Mongol conquests shifted into high gear after one particular incident. In 1218, the Mongol leader, Genghis Khan, sent a caravan to establish trade relations with the Muslim kingdoms of Central Asia (today's Kazakhstan and Uzbekistan). However, the caravan was slaughtered on the order of one of the local governors. We're not sure why he chose to murder a peaceful delegation, but it was possibly intended as a show of force. In retrospect, this was perhaps the worst decision in all of human history. Upon hearing the news, Genghis Khan carefully prepared an invasion, and within three years had razed the glorious cities of Central Asia to the ground, killing half the region's population.‡ Once unleashed, the horde

* Rice and wheat are both types of grass. Including what is sourced from grain crops around the world (rice, wheat, corn, barley, etc.), the human diet is collectively about 80 percent grass by caloric intake.

† The root of our word "undermine."

‡ It's estimated that around one in ten inhabitants of Central Asia today is a direct descendant of Genghis Khan, which implies a lot of wives and a lot of raping. There surely was a lot of both, but the one-in-ten figure may not be as

couldn't be contained. Subsequent campaigns by the descendants of Genghis Khan extended Mongol dominion across the Middle East as far as Egypt, across Russia into Europe, and into China, where they established the Mongol-led Yuan dynasty (1271–1368).

The Mongol invasions witnessed the first clashes between Europeans and the Far East. European knights faced off against Asian horsemen as the Mongols overran Russia and then moved into Central Europe. In 1241, a force of Mongols defeated a Hungarian army on the banks of the Sajo River and then pressed on to crush a force of European knights in full battle armor at Liegnitz in western Poland. Interestingly, Chinese contingents in the Mongol army may have used firearms in these battles, representing the first use of gunpowder in Europe. The Mongol army then advanced into Bohemia (the Czech Republic) and Austria, approaching the gates of Vienna—the capital of the most powerful European realm, the Holy Roman Empire. It's doubtful whether anything Europe might have done would have stopped the Mongol invasion. Instead, luck spared the continent from destruction. In December 1241, the sudden death of Ögedei Khan (Genghis's son) brought the invasion to a halt, as all the princes of Genghis's line were recalled to Mongolia to select a new leader.

The Middle East wasn't so lucky. By 1258, Mongols stood before the gates of Baghdad, center of the Muslim world. After a short siege, the city's defenses crumbled, and a week of pillage and plunder began. Palaces, mosques, libraries, and other public buildings were demolished. We may never know what knowledge was lost— perhaps the last copies of ancient Greek works that had survived the collapse of classical civilization by almost a thousand years. The city was depopulated, with hundreds of thousands killed. After Baghdad's sacking, the Mongols continued west to conquer Syria, until their expansion was finally stopped in 1260 by defeat at the Battle

---

significant as it seems. Simply by virtue of the doubling that occurs with each generation, each person's direct ancestors in the last two millennia easily outnumber all humans who've ever lived. (Of course, this relies on counting the same ancestor multiple times when they appear in multiple places on the same family tree.) In the same way, most people of European ancestry descend from Charlemagne.

of Ain Jalut at the hands of a Mamluk army from Egypt, which prevailed by using the same hit-and-run tactics as the Mongols.

The Mongols are remembered for their brutality, a reputation well earned. However, they can't simply be written off as purveyors of mindless destruction. For one thing, they played a major role in binding the world together. For another, the Mongols introduced a wide range of progressive ideas in the wake of their destruction. Once the initial phase of terror subsided, the Mongols could prove to be remarkably benevolent rulers, especially if a population had submitted easily. These feared horsemen of the steppes displayed curiosity, openness to new ideas, and religious tolerance.* They exempted the poor from taxation and encouraged literacy. Although they came from a simple nomadic background, they respected technical experts and would typically spare craftsmen and scholars. Like the Romans, the Mongols succeeded thanks to their hands-off approach; they allowed local populations to govern their own affairs as long as they paid tribute and kept the peace.†

To communicate across the largest terrestrial empire of all time, the Mongols developed a sophisticated postal service. A series of well-stocked posts allowed traveling messengers to exchange horses on the move. A messenger would typically gallop 25 miles between stations and either mount a fresh horse to keep going or relay the mail to the next rider to ensure the speediest possible delivery. Mongol riders could travel 125 miles a day, meaning that a message could get from Europe to China in less than a month. In some northern regions of Siberia, they substituted dogsled teams for horses. The Mongols also promoted trade, encouraging

---

* The Mongols originally practiced shamanism but soon converted to a wide variety of faiths, especially Islam, Christianity, and Buddhism. They even sponsored interfaith debates, inviting representatives from all over to gather and discuss philosophical and moral issues. Genghis Khan is thought to have enacted the world's first decree of religious freedom: a clear separation of church and state.

† This is how Moscow became the capital of Russia. Accorded the status of Mongol tax collector, the city was charged with collecting tribute from other Russian cities. Beginning as a minor town around 1147, Moscow became rich by siphoning off some of the taxes, so that by 1480 it had assembled enough economic and military strength to simultaneously overthrow the Mongols and assert its authority over other Russian cities.

merchants to visit their lands with guaranteed protection. Within their vast empire, the Mongols made a concerted effort to stamp out bandits, making trade routes safer than ever before. The Silk Road linking the Mediterranean to China, first established by the Greeks in the wake of Alexander, experienced a major revival under the Mongols. For the first time in history, it was possible to travel from Europe to China through a single realm, one that was well-disposed to foreign merchants.

This period when ideas, technologies, and goods flowed on a massive scale between Europe and Asia is called the Pax Mongolica ("Mongol Peace"). Europe's knowledge of the world vastly expanded thanks to information brought back by travelers and merchants. When Columbus sailed in 1492, his mission was to reach China to present a letter from Ferdinand and Isabella to the great Mongol khan, who they assumed was still in charge. The exchange went both ways. In 1285, Kublai Khan* ordered the creation of a single unified world map. Though lost to history, the resulting map was probably similar to the Korean Kangnido map of 1402, a copy of which has survived. The Kangnido includes details of not only Asia but Europe and Africa, too. Meanwhile, in Europe, the Italian cartographer Fra Mauro assembled the most comprehensive world map prior to the Age of Exploration, depicting the entire Old World in great detail. For the first time, Europeans and Asians had pooled their knowledge to assemble a complete understanding of the world as they knew it.†

The Pax Mongolica laid the groundwork for Marco Polo's travels. Although by 1271 the Silk Road had existed for over a thousand years, it mostly functioned as a bucket brigade. Goods moved from Asia to Europe, passing through dozens of hands, increasing in price as each middleman extracted a share. People rarely traveled the entire way. Marco was not the first European

---

* Ruler of China and grandson of Genghis Khan—"khan" just means "ruler" in Central Asia.

† The connection between East and West also came with some unexpected drawbacks. The Black Death, which struck Europe in 1347, was picked up by Genoese merchants in the Black Sea port of Kaffa, where it had traveled along the Silk Road all the way from China.

to travel to China, nor was he even the first in his family—he was accompanied by his father and uncle, Niccolò and Maffeo Polo, who'd been to China a decade earlier. What made Marco special was that he rekindled interest in the Far East with a bestselling book describing his travels, one of the most popular books before the printing press.* Some of Marco's tales are embellished, but there are enough details that we can be confident of its essential veracity. *The Travels of Marco Polo* not only brought knowledge of the Far East to readers back home, it inspired future travelers and explorers to go themselves. Columbus kept a copy with him and even consulted it as a reference guide to try to work out which part of Asia he'd reached (of course, this wasn't very helpful considering he landed in an entirely different continent).

At the time, Venice was the wealthiest city in Europe, a republic of prosperous merchants. Niccolò and Maffeo Polo had set out in 1261 on an expedition to Central Asia. There they met an emissary of Kublai, who invited them to China, since the great khan had never met a European. In 1266, they arrived at the imperial court in Beijing, where Kublai bombarded them with questions about European legal, political, economic, and religious systems. He then asked them to deliver a letter to the Pope (the closest thing to a "leader" of Europe) and return with a hundred European experts acquainted with the seven arts of grammar, rhetoric, logic, geometry, arithmetic, music, and astronomy. Also requested: a lamp with oil from the Holy Sepulcher in Jerusalem. Upon their return to Europe, the Polos were able to acquire the lamp, but instead of a hundred experts, they seem only to have been able to recruit the seventeen-year-old Marco. So in 1271, Niccolò, Maffeo, and Marco Polo set off from Venice for Asia.†

After sailing across the Mediterranean to Syria, the Polos rode

---

* Marco didn't write the book himself: he dictated it to a fellow prisoner after being captured in a sea battle against the rival Genoese, a few years following his return from Asia.

† This might be the first time Marco met his father, who had been away traveling in Asia for most of his life. It therefore shows a surprising degree of intrepidity on the part of Marco to suddenly leave his home to travel the world with virtual strangers.

camels to Hormuz on the Persian Gulf. From there, they intended to take a ship to China, but seeing the poor condition of the ships in port, they opted instead to continue overland along the Silk Road. This was a difficult journey across vast plains, icy mountain ranges, and some of the most hostile deserts on Earth. At one point they were ambushed by bandits after a sandstorm and barely managed to escape. Eventually, three and a half years after leaving Venice, the Polos were presented to Kublai at the imperial court in Beijing.* The khan was especially impressed with young Marco, who became a court favorite. Marco was clearly intelligent and could speak at least four languages, but interestingly, he seems never to have learned much Chinese. He probably communicated with Kublai in Persian, which was commonly spoken by the Mongol rulers of China and was a sort of universal language across their empire.†

Kublai made Marco an official of the Yuan dynasty and gave him various assignments, which allowed him to travel around China. He visited China's southern and eastern provinces, as far away as Burma. He also described Korea, Japan, and Indonesia (firsts for a European), although it's unlikely that he actually visited these places. A striking thing about the accounts of Marco Polo is their lack of bias. As merchants, the Polos would have been accustomed to foreigners, but contemporaries might have criticized religious and cultural practices different from their own. On the contrary, Marco expressed great admiration for China, with its magnificent history and orderly society, describing it as a place Europeans could learn from. This evenhanded representation of foreigners has led to a characterization of Marco Polo as the first modern secular humanist.

* A measure of how the world has gotten smaller: today, flights connect Venice to Beijing on a daily basis, appropriately taking off from the Marco Polo international airport.

† Based on the words and descriptions in his book, historians assume that Marco learned Persian, Arabic, Turkish, and possibly some Mongolian, in addition to his Venetian dialect of Italian and probably some other European languages as well. If this seems like a lot, Pope John Paul II is reckoned to have been fluent in at least eight languages, with some ability in many more.

Marco became so important in the imperial court that it became difficult for him to leave. Kublai initially declined all the Polos' requests to return to Venice. Meanwhile, the Polos worried that if the khan suddenly died, his enemies might turn on them in the ensuing power struggle. Their opportunity finally came around 1292, when the ruler of Persia (Kublai's great-nephew) sent an emissary to China to find a potential wife. Kublai gave the Polos permission to sail with the wedding party to Persia and then make their way home from there. They sailed with a fleet of fourteen junks past Singapore, eventually reaching Hormuz on the Persian Gulf. From there they traveled overland to the Black Sea and thence on to Venice, twenty-four years and fifteen thousand miles after they set out. Marco had spent more than half his life abroad, not only visiting the Far East but deeply immersing himself in its culture. Since he'd left Venice at age seventeen and returned at forty-one, most of his friends wouldn't have recognized him.*

There were other European travelers, though none so influential as Marco Polo. One who came close was probably fictional. *The Travels of Sir John Mandeville* was a collection of tales that was even more fantastic than Marco's. The account, which was widely circulated in the mid-to-late fourteenth century, described a world filled with monsters; people with dogs' heads; the land of the Amazons, ruled by fierce warrior women; a giant bird that ate elephants; fields of growing diamonds; forests of pepper trees; a fountain of youth; and a bird in Egypt, the phoenix, that immolated itself to be reborn every five hundred years.

These tales strike us as obvious fantasies today, but at the time the book was taken as literal truth by Europeans who expected to find marvels around the world. Indeed, Columbus carried with him on his voyage to the New World, in addition to *The Travels of Marco Polo*, a copy of Mandeville's book. Mandeville, purportedly

---

* Marco had a tough time convincing Europeans that some wonders he saw were real, such as the giant shipyards of Hangzhou, technologies such as coal and gunpowder, or bizarre Eastern customs such as bathing on a daily basis. He did have at least some physical evidence of his journey: a Mongol servant named Peter returned with the Polos and continued to serve him until Marco's death in 1324. Peter must have had some interesting stories of his own!

an English knight, probably never existed—the book is thought to have been written by a French author based on secondhand accounts and outright fabrications. In this sense, it should be taken as not a trailblazing work of geography but instead an early work of travel fiction, whose value lay not in conveying information but in encouraging interest in exploration.

One real European who did travel to the Far East was a priest named Giovanni da Pian del Carpine, who led a Catholic delegation to the Mongolian capital a few decades before Marco Polo's journey. The delegates were well received, but they resoundingly failed in their mission to convert the Mongols to Christianity. A few years later, the adventurer-priest William of Rubruck, acting on behalf of the French king, failed in a similar mission. There was also at least one reverse Marco Polo, a Far Eastern traveler who visited Europe around this time. Rabban Bar Sauma was a Chinese (or possibly Turkic) monk who followed the same overland route as Marco Polo, but in the opposite direction, to immerse himself in European culture (communicating via an Italian interpreter in Persian). His written account gives a fascinating description of medieval Europe from the perspective of an Asian traveler, east looking west.

The Old World had been connected, and it was clear to both East and West which civilization was superior. If you embarked on a world tour in the year 1400 and were asked which was most likely to dominate over the next few centuries, you'd have to say China. As the Ming dynasty rose from the ashes of Mongol collapse in the late 1300s, China stood tall on the world stage. Ruling from the newly constructed Forbidden City in Beijing, China had twice as many people as Europe, each at least as wealthy as their European counterparts. The 1,100-mile-long Grand Canal was renovated as a transportation highway teeming with more vessels than all the navies of Europe. Imperial granaries allowed the population to be fed even during bad harvests, and reservoirs provided irrigation and a steady source of drinking water. Luxurious silks and porcelain fetched a high price around the world, allowing wealthy Chinese to live in a state of splendor that their European counterparts could only dream of. The Chinese bureaucracy was expanded to include almost half a million officials, and competitive entrance exams ensured strong literacy and social mobility unimaginable anywhere else.*

In stark contrast to the opulence of China, Europe at the time was a miserable backwater still recovering from the ravages of the Black Death, which had recently reduced the population by a third. Europe was a patchwork of hundreds of quarreling states divided by incessant warfare. Germany and Italy were fractured into dozens of small kingdoms, electorates, duchies, margraviates, republics, bishoprics, city-states, and just about every other political arrangement ever conceived. England and France were locked in the death throes of the Hundred Years' War. Spain was fragmented and still facing off against the Moors in Granada. Only the Italian merchant republics, such as Venice and Genoa, seemed to offer any prospect of hope for the future, but these were tiny, divided, and constantly squabbling over their share of a limited economic pie. Europe was theoretically unified under the Pope, but this rarely brought any kind of cooperation and merely ensured the perpetuation of a stifling religious dogmatism. Changes were on the horizon, thanks to the rediscovery of classical manuscripts preserved in Byzantine and Arab archives, but the Renaissance was still very much a work in progress.

To an outside observer, the notion that within a few centuries an isolated, impoverished, and backward Europe would come to dominate the world would have seemed fanciful in the extreme. In 1400, the continent accounted for a mere 10 percent of the world's land surface and 15 percent of its population. The largest city in the world was Beijing, with a population of almost a million. Of the ten largest cities in the world, only one—Paris—was European, but it was hardly the glittering metropolis it is today. Even Tenochtitlán, capital of the Aztecs, was probably larger. European cities were dark and dirty, with narrow streets and abysmal sanitation. Waste and corpses were dumped unceremoniously into the

---

* The Chinese examination system is the basis of Western public service entrance exams. It was adopted by the East India Company, and Britain, France, Germany, and the United States followed suit by the late 1800s.

streets and rivers. A dearth of street lighting made venturing out after dark a risky proposition. With a violent crime rate fifty times higher than modern-day America's, the entire continent must have seemed one giant combat zone. Public executions were common, and disease was rampant.

In the year 1400, China contained four of the world's ten largest cities. The river valleys of China were among the earliest places to develop civilization, and China is perhaps unique in maintaining a relatively consistent culture and political unification for more than two millennia. (The closest Western parallel would be if the Roman Empire were still around, ruling a unified Latin-speaking Europe from its ancient capital of Rome.) China has nearly always been the world's largest country, in terms of both population and area,* but it has also been especially innovative. A partial list of Chinese inventions includes, in alphabetical order, acupuncture, bells, belt drives, coffins, compasses, crossbows, dental fillings, dominoes, ephedrine, fireworks, flush toilets, forensics, gasoline, goldfish, horse collars, hot-air balloons, hydraulics, incense, inoculation, iron smelting, kites, lacquer, matches, moldboard plows, nail polish, napkins, negative numbers, noodles, paper, paper money,† playing cards, porcelain, printing presses, rudders, running water, silk, speedometers, stirrups, tea, tofu, toilet paper, toothbrushes, umbrellas, and wrapping paper. Some Chinese inventions from over a thousand years ago sound downright modern. The invention of gunpowder during the ninth century spun off into military technologies such as mines, cannons, exploding projectiles, and multistage rockets that flew using streamlined wings. Even some things we think of as quintessentially European may have been invented in China. Recent archaeological digs have discovered evidence of a low-alcohol fermented grain beverage consumed in

* Russia passed China in land area about three hundred fifty years ago (taking some territory from China in the process), but most of Siberia is very sparsely populated. Canada is also larger than China in land area, but only if you count the surface area of lakes—Canada is nearly 10 percent lake surface, while China is fairly dry with only a few large lakes.

† Paper money in China bore the inscription "All Counterfeiters Will Be Decapitated." So much more imaginative than "In God We Trust."

China since ancient times. In other words, the Chinese may have invented beer.*

Several important Chinese inventions relate directly to exploration. Sternpost rudders appeared around two thousand years ago, providing a more effective means of steering a ship (recall that Vikings used a giant "steering board" oar, hence "starboard"). Chinese vessels were vastly more durable than European designs, divided as they were into separate compartments that could remain watertight even if another part of the hull sprung a leak. Also invented in China were paddleboats—vessels that could forgo protruding oars or sails in order to navigate narrow river passages. Perhaps most important of all, the Chinese invention of magnetic compasses revolutionized navigation. For the first time in history, sailors could now accurately discern their ship's direction in any condition and without an external frame of reference.

At the start of the Age of Exploration, China was far more advanced than Europe. When Vasco da Gama led the first Portuguese expedition to India in 1498, the Indian Ocean was teeming with local sailors who had plied the sea lanes for thousands of years. The ensuing power struggle lasted a century, and the Portuguese nearly lost it. Yet, despite the intensity of the contest, the Portuguese were actually fortunate to have arrived during an extreme power vacuum. A mere half century earlier, unbeknownst to Europe, China had withdrawn from the region. As it was, Portuguese fleets were almost destroyed at the hands of Indian and Arab navies. Had the Portuguese instead encountered the magnificent Chinese armadas that had been making frequent voyages to India in the previous centuries, European domination of the region would never have been a prospect.

In the early 1400s, China operated the largest and most advanced navy in the world. China, far more than Europe, was poised to discover, settle, and conquer the world. Small divergences in history

---

* This is ironic because the Europeans who settled in China found it distinctly lacking in beer, and so established the most popular "Chinese" beers known today. Tsingtao Beer, for example, was established by Germans in 1903 at the city bearing its name, which at the time was the German equivalent of British Hong Kong.

could have witnessed a Chinese expedition to Europe to impose trade concessions favorable to the emperor, shutting down European attempts to reach Asia by sailing down the coast of Africa or across the Atlantic. Instead, it might have been Chinese explorers making geographical errors that led them across the Pacific in search of a direct route to Europe, stumbling into the Americas in the process. Instead of Spanish conquistadors, we might have had Chinese ones razing Aztec and Inca cities, feverishly seeking gold and silver. In this alternate history, Chinese navigators might have searched the Americas for a sea passage to Europe, with Chinese pilgrims founding colonies in California to escape the authority of imperial power. If history had taken only a slightly different course, Chinese might have been the predominant language in the Americas, and, consequently, the world's international language.*

During China's Age of Exploration, all this seemed a distinct possibility. From 1405 to 1433, during the early Ming dynasty, Chinese fleets commanded by Admiral Zheng He ventured into the Indian Ocean seven times. These were vast enterprises, as much to showcase the power of the emperor as to conduct trade. The 1405 expedition contained over sixty grand "treasure ships" accompanied by two hundred smaller support vessels, manned by thirty thousand sailors. Contemporary sources describe the largest ships as four-decked monsters over 400 feet long and 170 feet wide, carrying a massive amount of cargo and up to one thousand passengers, powered by sails on nine masts. This would place them among the largest wooden vessels in history, surpassing even the majestic European ships of the line built centuries later. By comparison, the largest of Columbus's ships, *Santa María*, was less than a sixth the size, and his entire fleet would have comfortably fit in the cargo hold of a single one of Zheng's ships.

---

* English is the de facto international language for two reasons: the economic and cultural influence of the United States, and the historical extent of the British Empire, which ruled a sixth of the world a century ago. Nowadays, a quarter of the people on Earth speak at least some English. In European countries without English as a first language (most of them), almost half the population can hold a conversation in English, although this varies by country from a low of 22 percent (Spain) to a high of 90 percent (the Netherlands).

Zheng He was not originally Chinese. Rather, he was the son of a Muslim official who served the Mongol Yuan dynasty in the province of Yunnan and was born with a different name. Taken prisoner in 1381, Zheng was ritually castrated to become a eunuch servant of Zhu Di, the future emperor of China. Over time, Zheng became a trusted favorite and personal friend. In a 1402 power struggle, Zhu Di stormed the imperial capital, deposing his nephew to take his place as emperor. Chinese emperors have historically changed their name upon ascent to the throne, and Zhu Di selected the title "Yongle," meaning "perpetual happiness." This event catapulted Zheng from his humble beginnings to a position as one of the leading figures of China. Awarded the title of grand director, he was handpicked to represent the emperor on a series of overseas voyages meant to demonstrate the power of the new Ming dynasty.

The scale of Zheng's expeditions was something entirely new in the world, with hundreds of ships manned by tens of thousands of sailors. Starting in 1405, Zheng's fleets traveled all over Southeast Asia, India, the Middle East, and East Africa. Along the way, he presented gifts of gold, silver, porcelain, and silk, and in return Zheng received ostriches, zebras, camels, ivory, and other local novelties. One especially prized gift was a pair of giraffes brought back from the Swahili Coast of Africa. Giraffes happened to match the description of a Chinese mythological creature called the *qilin*, thought to be an omen of good luck (analogous to the European concept of unicorns). These giraffes were the centerpiece of the Beijing zoo for many years, interpreted as a blessing on the emperor's rule. Zheng's voyages established diplomatic contact between China and distant lands, with over thirty rulers from around the Indian Ocean sending emissaries to pay homage at the Chinese imperial court.

Although Zheng's ships were sailing along routes Chinese merchants had known for centuries, this was the first time that China came in force. The fleet was armed with hundreds of bronze cannons, enough to make short work of any Portuguese incursions into the Indian Ocean, had the timing coincided. However, Zheng usually attained his goals through diplomacy rather than military

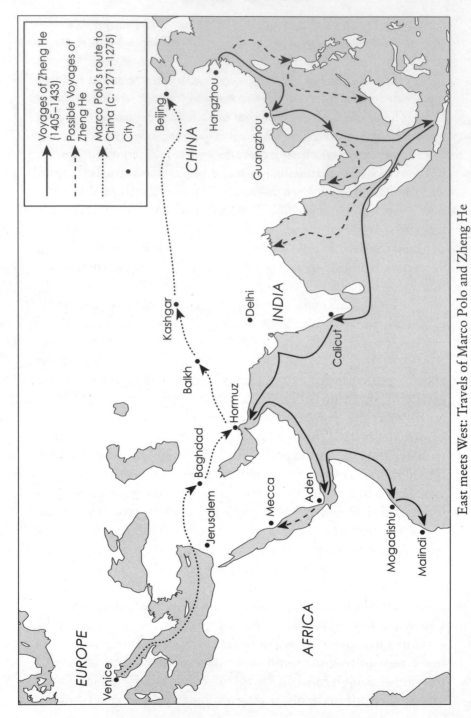

East meets West: Travels of Marco Polo and Zheng He

Legend:
- Voyages of Zheng He (1405–1433)
- Possible Voyages of Zheng He
- Marco Polo's route to China (c. 1271–1275)
- City

Labels: EUROPE, Venice, Jerusalem, Mecca, Baghdad, Balkh, Kashgar, Hormuz, Aden, Mogadishu, Malindi, AFRICA, Delhi, INDIA, Calicut, CHINA, Beijing, Hangzhou, Guangzhou

force. Merely the sight of his powerful navy was enough to awe potential adversaries into submission, and the fleet's weapons were rarely used in combat. However, Zheng didn't hesitate to arrange threatening demonstrations on several occasions, one time deciding to depose a ruler of Ceylon to replace him with one more favorable to the Chinese. He also took the opportunity to exterminate pirates who'd long plagued Asian waters.

The voyages had a long legacy. An overseas diaspora resulted in millions of Chinese moving to places such as Malaysia, Indonesia, and the Philippines, spreading Chinese technology and culture as immigrants mixed with locals. This is why Chinese is an official language of Singapore today and is commonly spoken throughout Southeast Asia. Zheng became a figure of folk veneration among expatriate Chinese, with temples built in his honor. Chinese silk, tea, porcelain, and lacquered furniture became common throughout the world, and would eventually spark a Far Eastern craze in artistic circles. Europeans of the seventeenth and eighteenth centuries wanted their houses and palaces stocked with Chinese styles, coveting, for example, blue-on-white patterned Ming vases and Chinese-inspired Dutch Delftware. As a result, we still call ceramic dinnerware "china."

In 1424, the Yongle emperor died. One final expedition was authorized in 1431, but after that, the great voyages of discovery ended. Zheng He perished shortly thereafter, and was probably buried at sea. In the following decades, imperial officials ended state-sponsored exploration, and most records of the voyages were lost, either deliberately or by neglect. The voyages were canceled to save money, but it was probably more than that. They represented a policy that promoted interactions with foreigners, exposing China to new ideas that could prove dangerous to those in power. As conservative Confucianists consolidated control, they implemented a policy of forced isolation, ending overseas contact. China stopped exploring.

China, the greatest nation on Earth, was assumed to be perfectly self-sufficient. Its vast canal system and internal economy should be enough to sustain a high degree of development. What use did the Celestial Empire have with foreigners and their ideas?

By 1500, it became a capital offense to build a seagoing junk with more than two masts. The ban on overseas trade and destruction of the Chinese fleet ruined local economies and sparked an explosion of smuggling and piracy in coastal communities that depended on the sea. Japanese pirates and Europeans moved in to fill the vacuum and reap profits from overseas trade that had once filled the coffers of the Chinese imperial court. Instead of a global economic leader, China became a victim of punitive treaties at the hands of predatory nations. Cut off from the outside world, Chinese technological development stagnated as Europe went on to occupy the commanding position on the global stage that China had once aspired to.

10 | A SEA ROUTE TO INDIA

Thousands of miles from the majestic imperial court in Beijing, on the far western edge of the world's largest landmass, Europe was slowly emerging from the Dark Ages. By contrast with China, Europe was poor, technologically backward, and politically divided. An alien visiting from space would hardly pick out Europe as likely to lead a new age of global exploration, but by the late 1400s there were glimmers of hope. The population was recovering from the ravages of the Black Death. In its wake, serfdom had finally ended, and wages and freedom of movement were on the rise. The classics of ancient Greece and Rome had recently been rediscovered by the republics of Northern Italy, where they were inspiring new forms of art and philosophy that were starting to push back the curtains of religious dogmatism. Europe wasn't modern yet—it would be centuries before the full force of the scientific revolution overturned society—but change was in the air.

By the late 1400s, the economic void left by the fall of Rome had been filled, with prosperity reaching levels equivalent to those at the empire's height. The Mediterranean world was once again connected to the north, this time with the full participation of the former "barbarians." The Baltic Sea became a hub of activity with the creation of the Hanseatic League, a flexible free-trade alliance of more than a hundred towns that grew from merchant guilds that imported amber, furs, and naval supplies from the hinterlands of Scandinavia and Russia. Each town was nominally independent,

125

but they often united in times of crisis. Ships from many towns would sail together for mutual protection from pirates and predatory states.*

This emerging prosperity in the north was mirrored in the Mediterranean. Trade routes of the Italian merchant republics linked with those of the Hanseatic League to create a continental commercial network for the first time since the fall of Rome. Venice and Genoa controlled much of the southern trade, with Venice completing the last link of the spice route from Egypt while Genoa focused on ancient links with Asia via the Black Sea. Meanwhile, Florence became the center of an emerging banking industry, with florins as the first international currency. Before banks, merchants had to carry cash or wares for exchange. Banks made commerce vastly easier and safer because cash deposited in one city could be withdrawn in another to purchase goods. They also provided loans and capital to finance new ventures, allowing for more ambitious expeditions than ever before.

The trade routes were major conduits of knowledge. Contact with the Middle East and Byzantine Empire brought the Renaissance home to Italy. The recovery of lost manuscripts preserved in the East reminded Europe of what it had once known. At the same time, Byzantine scholars fled to Italy as the Ottomans encroached on their empire, bringing an influx of linguistic experts to teach at emerging universities. Humanist scholars pored through the archives of monastic libraries to see what still remained of ancient works. European art had for a thousand years been restricted to drab religious icons. The rediscovery of the classics sparked a surge of humanist optimism: a drive to emulate and then surpass the ancients. Curiosity had returned to European shores. The Renaissance was accompanied by urbanization, as labor shifted from agriculture to commerce and specialist trades. Though still tiny compared with Asian metropolises, Venice, Florence, Genoa,

---

* "Hansa" is from an old German word for "guild." The Hanseatic League was one of the inspirations for the European Union and lent its name to the German airline Lufthansa ("Air Guild"). Hamburg, Lübeck, and Bremen are still classified as "Free Hanseatic Cities" within the German Federation (not part of any state).

and Milan grew from small towns into major cities of over one hundred thousand inhabitants.

The expansion of trade in Europe drove the evolution of ship design. First to come were cogs, sturdy oaken ships developed for the Baltic and North Sea trade of the Hanseatic League, with wooden planks held together with iron spikes and sealed watertight with tar. Cogs mounted a single mast with square sails and were rather stout, with rounded hulls and high castles in the bow and stern. Pretty much your basic medieval sailing ship, cogs weren't much to look at, being ungainly and slow. But with a respectable cargo capacity of up to two hundred tons—a bit more than Columbus's largest ship, *Santa María*—cogs were the workhorses of European trade for five hundred years from the eleventh century. They were even pressed into service as warships, with archers mounted in the bow and stern castles, although medieval naval combat mainly consisted of crashing your ship into the enemy's and fighting hand to hand.

Gradually, and especially after Europeans began sailing to Asia and the Americas, cogs evolved into carracks. Originally developed by the Spanish and Portuguese, carracks were a fusion design combining the best characteristics from Northern Europe and the Mediterranean. Specifically designed for the rough Atlantic, carracks were more stable than cogs in high seas, introducing a more streamlined shape and improvements in sails and rigging. They usually carried square sails for speed in favorable winds, supplemented by a triangular sail at the rear for maneuverability. Carracks maintained the high bow and stern castles, but these were less pronounced than on cogs. Columbus's *Santa María* was a small carrack. Carracks were the archetypical European sailing ship that set the standard until wood and sails were replaced by iron and steam. Galleons, sloops, frigates, and ships of the line were all variants of the basic carrack design.

Caravels were a smaller ship design introduced by the Portuguese for long-distance exploration down the African coast. With a shallow draft, caravels were highly maneuverable, capable of sailing upriver and into shallow coastal waters. Triangular sails—a recent introduction to Europe inspired by the dhows of the Indian

Ocean—meant that caravels could effectively sail into the wind, although they could also be equipped with square sails for speed in favorable winds. Thanks to their versatility, caravels became mainstays* of early European exploration. Columbus took two with him to the New World (*Niña* and *Pinta*), and caravels were critical in opening up the Indian Ocean spice trade for the Portuguese. Later, caravels, with their limited cargo capacity, were mostly supplanted by larger carracks and galleons.

Portugal led the European Age of Exploration. A small kingdom on the far western edge of Europe with few resources, Portugal did at least have easy access to the Atlantic and was strategically placed at the nexus of northern and Mediterranean trade. Mariners from north and south would stop in Portugal to provision their ships and exchange goods. Originally a Spanish province,† Portugal was declared independent by an ambitious prince in 1139 at the old Roman harbor town of Portus Cale, which would eventually lend its name to the new nation.‡ This city is today called Porto, the second-largest city in the country, renowned for its "port," fortified wine. In 1249, Portugal expelled the Moors from the southern part of the kingdom and moved the capital to the newly conquered port of Lisbon, igniting a rivalry between the country's two leading cities that continues to this day. (Lisbon, sitting on one of the best natural harbors in Europe, was also an important Roman city, and before that a Phoenician trading post dating back thousands of years.)

Portugal was spurred to action by its energetic prince Henry. Known to history as "the Navigator" (even though he never embarked on voyages of discovery himself), Henry was an enthusiastic promoter and patron of the overseas expansion of Portugal. The third son of John I of Portugal, Henry never became king,

* A nautical term meaning the main support for a ship's mast.
† Spain consisted of multiple small kingdoms until it was united by Ferdinand and Isabella. Today, part of Spain (Catalonia) wants to reverse the unification and become independent once again.
‡ "*Portus*" means "port" in Latin. "Cale" predates Rome, probably deriving from the Celtic for "port" ("*cala*"), meaning the town was named "Port Port" in a combination of Latin and Celtic.

but he nevertheless exerted a major influence on Portuguese foreign policy. He recognized that as a small agrarian country with a single dominating neighbor to trade with (Spain), Portugal must look across the ocean for its advancement. In 1415, Henry encouraged his father to seize Ceuta, a Muslim port in Morocco opposite Gibraltar.* This gave Portugal access to the caravan routes that regularly brought gold, ivory, slaves, and exotic treasures across the Sahara from the wealthy kingdoms of Africa. Henry was intrigued by Africa and its riches, and wondered what lay on the other side of the desert—no European had traveled into the continent or down its coast since Roman times.

Soon after the conquest of Ceuta, Henry dispatched a Portuguese knight named João Gonçalves Zarco to sail down the coast of Africa. He was blown off course, only to find shelter at an island the sailors called Porto Santo. This turned out to be a serendipitous discovery, so Henry claimed Porto Santo and its larger companion Madeira as Portugal's first overseas colonies in 1420. These islands were known to the Greeks and Romans, but to medieval Europeans they were new lands. Madeira became a major Atlantic hub and the first European colonial experiment. The climate suited sugarcane, whose labor-intensive production marked the beginning of the slave trade when a shipload of captive Africans arrived in 1452. The European appetite for sugar soon grew too fast for Madeira to satisfy, so the plantations were relocated to Brazil while Madeira shifted to producing the wine it's known for today.

Meanwhile, Portuguese navigators continued their voyages into the Atlantic, and in 1427 discovered the Azores. These islands lie more than twice as far out to sea as Madeira: at over nine hundred miles from Portugal, they're fully a third of the way across the Atlantic. Indeed, the Azores are the closest thing to a stepping-stone between Europe and the Americas, which made them an important base for future sea voyages (and later an important air base at the dawn of intercontinental air travel and during the Second World War). Columbus stopped at the islands on his way back

---

* Later Ceuta passed to Spain, under whose control it remains: a tiny European enclave on the continent of Africa.

from the New World to repair his storm-beaten ships. Although interestingly, the Portuguese governor of the Azores at the time of his arrival may actually have had a better claim to have discovered the Americas than Columbus did.

In a history of the islands published in the 1570s, the Portuguese priest Gaspar Frutuoso claimed that their governor, João Corte-Real, had been awarded his post for discovering "Terra Nova do Bacalhau" ("New Land of the Codfish") across the Atlantic in 1473. This is not impossible: Basque fishermen are known to have sailed to the Grand Banks off the coast of Newfoundland around this time, and islands weren't only suspected in the Atlantic but actually marked on contemporary maps. This kind of sighting might not have been major news, and since navigational charts were kept as state secrets by the Portuguese, the news wouldn't have traveled far. However, since the governorship grant merely mentioned vague "services rendered to the Portuguese Crown" and the claim was made almost a century later, we must approach it with a great deal of skepticism.

Shortly after the discovery of the Azores, Prince Henry redoubled his efforts to explore down the coast of Africa. This required sailing past the treacherous Cape Bojador near the westerly bulge of the continent, whose portent is conveyed by its Arabic name "Abu Khatar" ("Father of Danger"). Shoals of sardines and waves crashing on shallow reefs make the sea appear to boil, stoking fears of monsters expected to inhabit the sea.* Stifling winds wafting from the Sahara heighten this impression of heat, while ferrous minerals cause compasses to veer in strange directions. The cape was still notorious for its strong currents and winds centuries later, with at least thirty shipwrecks between 1790 and 1806. Eventually, Portuguese navigators learned that hugging the coast was simply too dangerous and changed their tactics by rounding the cape far out to sea.

---

* It may seem naive to imagine sea monsters, but in a way the fear was justified. Medieval sailors wouldn't have known the particulars of marine life, but twenty-five-foot-long flesh-eating sharks, sixty-foot-long giant squid, and hundred-foot-long whales seem to fit the bill, not to mention the giant sharks and reptiles that have inhabited our oceans in times past. Sea monsters do exist—superstitious sailors were merely wrong about the details.

By 1444, the Portuguese reached the Senegal River south of the Sahara. With this water route into the continent's interior, Europeans bypassed caravan routes that had made West African kingdoms wealthy for centuries. Gold, ivory, and other exotic goods of the tropics began to pour into Lisbon by the shipload. Each year, dozens of vessels sailed to Africa on behalf of the Crown, and soon private merchants began organizing their own expeditions. We often think of the Age of Exploration as an attempt to circumvent Muslim control of the Silk Road by sailing around Africa to reach India. This was certainly an objective, but there were also powerful kingdoms in Africa to trade with, filled with luxuries Europeans could only imagine. In the 1300s, the Mali Empire was the second largest in the world after only the Mongol Empire, supporting a population a third the size of Europe's. Its majestic metropolis of Timbuktu was one of the richest cities on Earth.* The subsequent Songhai Empire (1464–1591) was even larger.

The Portuguese, after all, didn't know if sailing around Africa was possible. They assumed it was, based on theories of ancient Greek geographers, but these might have been conjecture. There was one report of ancient ships sailing around Africa—the Phoenician-Egyptian expedition recounted by Herodotus—but the Portuguese would have been as skeptical of this story as we are. As they pressed south, mile after mile, encountering nothing but more coastline, they must have had their doubts. Nevertheless, even the West African trade was a major boon to the Portuguese economy, and anyway, the Portuguese understanding of the origin of spices was rather hazy. Wherever they went in Africa, they took samples of spices to show the locals and inquire if any grew nearby, receiving only bemused looks in return.

Meanwhile, Portugal claimed vast stretches of African territory by raising wooden crosses bearing the royal insignia. Set-

---

* We previously met Mansa Musa of Timbuktu, the richest person in history. Most people don't realize that Timbuktu is an actual place, but rather use it as a byword to convey something like "the middle of nowhere." It's ironic that the city probably earned this reputation by being the best-known place in sub-Saharan Africa, a place that most Europeans had heard of but associated with remoteness.

ting aside the fact that people already inhabited these lands, the rules of European land claim seem decidedly arbitrary to modern eyes. Even if divine authority to appropriate territory somehow flowed from the pontiff in Rome, how far did this power extend? This was a question faced by the early American colonies. When first proclaimed, they extended indefinitely inland. In other words, the original thirteen American colonies technically stretched all the way to the Pacific Ocean. This proved impossible to maintain, so new states haphazardly split off from old ones: Vermont from New York, Kentucky from Virginia, and Tennessee from North Carolina. How much territory could an explorer claim? When Vasco Núñez de Balboa reached the Pacific Ocean for the first time in 1513, he waded knee-deep into the surf with a sword in one hand and a standard of the Virgin Mary in the other, claiming possession of the ocean and all adjoining land for Spain. The Pacific covers 46 percent of the world's surface, bordering five continents. Claiming half the world in some kind of solemn Lady of the Lake ritual seems more than a tad ridiculous.

When Columbus returned from the New World, the Spanish and Portuguese (the only players in the game at that time) tried to solve this problem with the Treaty of Tordesillas in 1494. This divided the entire non-Christian world between them, along a north-south line running through the Atlantic, midway between the Portuguese Cape Verde Islands and the Caribbean islands discovered by Columbus. Cuba was named in the treaty as "Cipangu" (Japan), reflecting how incredibly little they knew about the world. Brazil was discovered six years later, a portion of the Americas jutting out across the Portuguese side of the line, which is why it was settled by Portugal and speaks Portuguese today.* In fact, the treaty was a massive gamble, because although the line theoreti-

---

* During Napoléon's invasion of 1808, the Portuguese king fled to Rio de Janeiro. Although he moved back to Portugal in 1821, his son Pedro remained as regent. Declaring independence the following year, Pedro began a line that ruled the Empire of Brazil until 1889. This gives Brazil two distinctions: 1) it's the only American country to have ruled a European one (Portugal, from 1808 to 1821), and 2) it's one of the only non–Native American monarchies to have existed in the Americas (Mexico also formed two brief empires).

cally extended around the globe, no one knew where it fell on the other side of the world. This meant that the valuable Spice Islands, whose location was still a mystery, could have belonged to either. Even once they were discovered, since navigators at the time had no way of measuring longitude, it was impossible to sort out in whose domain they lay. Portugal ended up with the Spice Islands because they arrived first, while Spain kept the Philippines for the same reason, even though both lie at precisely the same longitude.

The idea of dividing the world between two tiny European powers strikes us today as not only supremely arrogant but utterly absurd. The treaty was ignored by not only Protestant powers, such as England and the Netherlands, but even Catholic France.* Anyway, apart from some remote islands far out at sea, the entire world had already been settled for thousands of years. Of course, this wasn't seen as a major impediment to Europeans. A typical British rule of thumb was if natives had a clear hierarchy with a ruler such as a king (i.e., someone to negotiate with) and practiced agriculture (i.e., they were using the land "properly"), then at least a pretense of a treaty must be signed to seize territory. This was, of course, subject to manipulation: locals could be bribed or coerced into signing away land they never owned—especially considering the entire concept of land ownership was quintessentially European. Land could often be purchased at bargain rates for wampum,† guns, or alcohol. A famous example was the Dutch purchase of Manhattan in 1626 for trade goods reckoned to be worth sixty guilders (around $24).‡

As they claimed territory and established trading posts, Portuguese navigators pressed ever farther down the coast of Africa. By 1482, Diogo Cão crossed the equator for the first time and sailed

---

* Although the 488-year-old Treaty of Tordesillaswas bizarrely invoked as one of the justifications for Argentina's invasion of the Falkland Islands in 1982.

† Seashells used as currency by Native Americans in the northeastern United States.

‡ The $24 estimate would be over $1,000 today adjusted for inflation. On the other hand, if you invested $24 from the purchase in 1626 at a rate of return of 7 percent, it would be worth over $5 trillion today: more than three times Manhattan's property value. So maybe it was the Dutch who were swindled?

his caravel up the Congo River into the heart of Africa. There, he established relations with the Kingdom of Kongo (distinct from modern Congo but lending its name to the region). Kongo was an African civilization based on the cultivation of yams, beans, sorghum, and millet, with a hilltop capital supporting a population in the tens of thousands. Around the size of California—five times larger than Portugal—Kongo projected its influence across Southern Africa. Kongo was a trading partner of Great Zimbabwe, through which it had connections to the Indian Ocean. At least one Portuguese sailor spent several years in Kongo, and later became the official Bantu* translator on Vasco da Gama's 1497 voyage to India.

In 1488, Bartolomeu Dias finally rounded the southern tip of Africa, sixty-eight years after the first voyage commissioned by Prince Henry the Navigator (who had died in 1460). Perhaps the longest sea voyage up to that time, Dias's sixteen-month journey now seems modest compared with some later expeditions. After sailing south along the west coast of Africa, Dias and his two caravels were pummeled by massive storms for thirteen days. When the storms finally cleared, they sailed east expecting to find land. Finding only open ocean, Dias turned northeast. A month later, he landed in Mossel Bay, at the southern tip of Africa. Dias continued sailing for another month to the point where the African continent turns north, opening up to the vast Indian Ocean. There, near Kwaaihoek, South Africa, they erected a monument to mark their farthest point. Dias wanted to sail on to India, but his crew refused to go any farther, so they sailed for home to report the news. It was only on the return journey that they discovered the Cape of Good Hope, which they named the "Cape of Storms." (The name was changed to "Good Hope" in a fifteenth-century PR campaign with echoes of Erik the Red's naming of Greenland.)

Shortly after Dias arrived in Portugal, a man named Christopher Columbus returned to Spain claiming that he'd reached the Far East. This was somewhat alarming to the Portuguese, although it was far from clear what he'd actually found—so with the route

---

* Bantu is one of the largest language groups in Southern Africa.

around Africa now open, the Portuguese moved quickly into the Indian Ocean. On July 8, 1497, Vasco da Gama sailed from Lisbon with four ships and a crew of 170 handpicked sailors. One of the reasons da Gama was chosen instead of Dias is that the Portuguese liked to rotate their captains so no single one became too famous or powerful. Thus, Dias was replaced by da Gama, and da Gama was later replaced by another captain. Vasco da Gama's voyage would be even longer and harder than Dias's had been, with half his ships and two-thirds of his crew lost during an ordeal lasting more than two years.

By December 1497, da Gama's ships were passing the point where Dias had turned back. With Christmas approaching, they named the coast "Natal" (for the nativity), which the region is still called. Around this time, the crew experienced the first symptoms of a disease that would become the bane of every seafarer of old: limbs swelling, gums bleeding, and teeth falling out.* Scurvy forced a stop at the Zambezi River in Mozambique, where they buried their dead and repaired their ships before continuing on. Soon they encountered Arabic speakers—a sign that they were nearing the Indian Ocean trade network. Sailing up the Swahili Coast, they arrived in the Kenyan port of Mombasa, its harbor full of Arab merchants. Here, the Portuguese met great hostility. At first, they assumed it to be religious in nature, but in fact, the Arab merchants must have been alarmed by the sudden appearance of Europeans for the first time in the Indian Ocean.

Vasco da Gama still needed to figure out how to get to India. Sailors in Mombasa offered to show him, but this turned out to be a ruse to wreck the Portuguese ships along some shallow reefs. Discovering the plot by torture, the Portuguese sought vengeance by bombarding the town with cannons and burning some ships in the harbor before continuing on. Sailing up the coast to the port of Malindi, they received a warmer welcome from the city, a rival of Mombasa. In Malindi, the Portuguese learned of the Christian

* It wasn't until 1934 that scurvy was traced to a deficiency of vitamin C. Recall that the Vikings had (probably accidentally) prevented scurvy by eating onions, radishes, and sauerkraut—far healthier than later diets of dried meat and "hardtack" biscuits.

kingdom of Ethiopia farther inland, and they wondered if this might be the legendary realm of Prester John, a potential ally that might be persuaded to come to the aid of Europe against the Muslims of the Middle East.* They also found a friendly navigator who could guide them the rest of the way. Thus, the Portuguese didn't so much discover the route to India as ask some locals for directions.

Taking advantage of monsoon winds, da Gama's ships crossed from Africa to India in less than a month, anchoring at Calicut on May 20, 1498. Not knowing how they'd be received, da Gama proceeded with caution. He put a single sailor in a boat to have him scope out the town. Landing at the docks, the sailor was shown to the foreign quarters, where he found Arab traders who demanded in perfect Spanish, "What are you doing here?" The traders had come from Tunis in North Africa and made it clear that the Portuguese were not welcome in the Indian Ocean. Later, exploring the city, the Portuguese sailor got lost in the back streets of Calicut and stumbled upon a Hindu temple. He was amazed to find that the temple contained images of "saints," some of whom had halos.† Since Muslims weren't permitted to display images of people, and since the sailor had never heard of Hinduism, he delightedly inferred that this was a Christian church. This assumption was reinforced by chants of "Krishna" (Hindu god of compassion and love), which the sailor assumed referred to "Christ." Yet, as he reported back to the fleet, the Indians worshipped such strange saints, some of whom had multiple arms and blue skin; one even had the head of an elephant!

After a few days, the Portuguese managed to secure an official audience with the city's raja. Da Gama presented gifts from the Portuguese king: sugar, oil, honey, cotton, coral, and ornamental washing bowls. Indian officials laughed derisively at the worthless collection of trash the so-called "mighty" king of Portugal had the impertinence to send. They wondered why there was no gold or silver, the only gifts suitable for an illustrious potentate. Once the

---

* Ethiopia practices an archaic form of Christianity dating to before the Muslim conquest of Egypt in 640, when it was cut off from Europe.

† Halos go back at least three millennia to Homer's *Iliad* but first appear in Asian art around the time Alexander arrived in India—likely a Greek influence.

Portuguese delegation left, Arab merchants arrived to confer with the raja, and they suggested that da Gama was no royal ambassador, but more likely a common pirate. India, one of the wealthiest lands on Earth, had no need for European trinkets: the bazaars of Calicut were filled with precious stones, pearls, silks, and valuable spices.

Eventually, da Gama was able to purchase a small cargo of pepper and cinnamon, but he realized that Arab influence in India would have to be broken by force. Leaving Calicut after three months, he ignored the advice of local sailors and decided to try to make the crossing to Africa against the monsoon winds. It was a harrowing journey. The outward crossing to India had taken a mere 23 days. The return voyage was more than five times as long—a grueling 132 days against the wind. Half the remaining crew was lost in the crossing, and most of the survivors were plagued by scurvy. Eventually, the fleet reached the African coast in January 1499, where it docked in Malindi to recuperate.

The rest of the voyage home was long, if relatively uneventful, although da Gama himself stopped in Cape Verde to tend to his brother Paulo, who'd fallen grievously ill during the voyage. Paulo soon died, and da Gama made his way home a month later on a caravel returning from West Africa. Despite the expedition's heavy toll, this first European voyage to India since Roman times was considered a resounding success. Da Gama was awarded the exalted title of "Admiral of the Seas of Arabia, Persia, India, and all the Orient," perhaps as a direct challenge to Columbus's Spanish title "Admiral of the Ocean Seas." The acquired spices were sold at enormous profit, paying for the expedition many times over. Yet, the voyage also had revealed the extent to which the Portuguese would have to commit military force to control the Indian Ocean.

Unlike da Gama's exploratory voyage, Portugal's follow-up expedition was decidedly military, led by Pedro Álvares Cabral with thirteen warships and fifteen hundred soldiers and sailors. Upon reaching India, Cabral constructed a fortress as an operational base at the small port city of Cochin. Next, he sailed for Calicut to establish a trading post. Hostilities ensued with Arab merchant guilds, and the Portuguese retaliated by ruthlessly bombarding the city with their cannons. Initiating a blockade, they started attacking all ships sailing to and from India—including Muslim pilgrim vessels bound for Mecca, whose passengers the Portuguese brutally massacred by sealing the ships and burning them to the waterline. Meanwhile, more Portuguese fleets arrived every year with more soldiers as the conflict escalated.

One night while the Portuguese fleet was anchored outside Calicut, they heard a splash and saw a swimmer coming toward them. This turned out to be the Italian adventurer Ludovico di Varthema, who, following in the footsteps of Marco Polo, had ventured across the known world. He was hoisted on board and supplied warm food and dry clothes, whereupon he described his incredible voyages. Varthema had become a mercenary soldier in service of the Mamluk rulers of Egypt, and from there had traveled overland to Persia while guarding a caravan. On the way, he'd had the opportunity to see Mecca and Medina, possibly making him the first Christian visitor to the holy cities of Islam. From Persia, Varthema took a ship to India and later traveled to the Far East, stopping at the trading center of Malacca and becoming perhaps the first European to visit the Spice Islands.

The Portuguese were fascinated to meet a fellow European in such a distant land. But although Varthema was certainly unusual, he was by no means unique. The fact that Europeans could sail halfway around the world and meet travelers from home was an early sign of how connected the world already was. Varthema became a key source of Portuguese intelligence on Asia. The first thing he did was warn of an impending attack. Ordering crews to battle stations, the Portuguese fleet used concentrated cannon fire to fight off an assault by two hundred small Arab and Indian ships at the 1506 Battle of Cannanore.* Varthema supplied the Portuguese with details on the defenses of Indian harbors, and his language skills were vital in negotiating with local rulers. Nevertheless, despite repeated attempts to take Calicut, the city man-

* One of the young Portuguese soldiers wounded in the battle was someone we'll meet later: Fernão de Magalhães (a.k.a. Ferdinand Magellan).

aged to hold out until 1588, when it finally succumbed to the Portuguese.*

As the struggle for India continued, the Portuguese moved on to conquer Malacca on the Malay Peninsula in 1511, to assert direct control over the spice trade. The goal was to establish a string of Portuguese bases stretching all the way from Europe to the Far East, while stifling competition by dismembering the Indian Ocean network that had existed for thousands of years. In 1513, the first Portuguese expedition to China dropped anchor at Macau, which became the longest-lasting European colony in Asia, persisting for almost five centuries until it was finally returned to China in 1999. By 1543, the Portuguese reached the outer limit of Asia when they landed in Japan, less than fifty years after Vasco da Gama had set out to find a sea route to India. The country on the far western edge of Eurasia was now connected with its counterpart on the opposite side of the globe. This completed the last link in the chain forever binding the Old World together into a single globalized system.

* Who were replaced by the Dutch, then British, then Dutch again, then French, then British again over the coming centuries until India gained independence in 1947.

11 | PLUNDER AND GOLD

As the Portuguese worked their way down the coast of Africa, a Genoese navigator named Cristoforo Colombo was formulating plans to reach Asia by a shorter route. He was not the first European to reach the Americas—the Vikings had been there centuries earlier, and others may have also—but Columbus was the first to initiate a sustained exchange between the Old World and New. For the first time in history, people, plants, animals, diseases, and ideas flooded across the Atlantic, and soon the Pacific, to bind our world into a global network of interconnectivity. Though the reputation of Columbus has suffered in recent years as we consider his legacy from a modern perspective, there's no doubt that he was an important historical figure. This is all the more fascinating because his importance lies in something that until his dying day he refused to admit he'd done. Even after four voyages, Columbus never accepted that he'd reached a new continent instead of Asia.

Christopher Columbus was born in 1451 into a middle-class Genoese family. His father was a wool and cheese merchant, and young Christopher gained experience in the business, going to sea at the age of ten. Genoa was a merchant republic and home to the most respected cartography school in Europe. Italian cartographers had recently innovated realistic navigational charts that used compass directions and sailing distances as references. Before these so-called portolan charts, most maps simply depicted the relative location of interesting places, often for religious purposes, without much thought about precision or scale. As they sailed down the African coast, Portuguese navigators consulted Genoese cartographers, and Columbus's brother Bartholomew had worked as a mapmaker for years. Columbus was born at an auspicious time and place for a future would-be explorer, with the Italian Renaissance blooming and printed books describing rediscovered works of science, geography, and astronomy becoming widespread during his youth.

In 1473, at the age of twenty-two, Columbus took a job as a traveling agent of the Genoese Centurione bank. We're not sure of the full extent of his travels, but he likely sailed all around the Mediterranean and possibly into the Black Sea, which hosted Genoese colonies at the time.* In 1476, Columbus sailed with a convoy to England, but the expedition was attacked by French pirates off the coast of Portugal. His ship destroyed, Columbus was forced to swim several miles to the Portuguese shore. Eventually he made his way to Lisbon, which, as we've seen, was a hive of mercantile activity. Its docks were thronged with sailors from all over Europe, and Portuguese ships were regularly returning with slaves, gold, and ivory from Africa. This surely would have made an impression. Columbus continued on with the local branch of the Centurione bank, and soon made his way to England and Ireland. In Ireland, he would have heard about St. Brendan the Navigator, who'd discovered new lands across the ocean.

In fact, it was common for contemporary European maps to show islands far out into the Atlantic. What made Columbus different was that he wrongly concluded that these islands were outlying outposts of Asia. According to a biography written by his son, Columbus's idea that the lands across the sea were populated was inspired by an event he witnessed in Ireland. One day a mysterious boat washed ashore with two exotic bodies. These could have been Native Americans (possibly Inuit) who drifted off course—prevailing winds would have carried them to Ireland,

---

* We saw these as Greek colonies visited by Pytheas. The largest, Kaffa, was the origin of the Black Death in Europe, introduced during a 1347 Mongol siege. The Ottomans conquered Kaffa in 1475, a year or two after Columbus would have visited.

and the corpses may have been preserved by the cool northern weather. Of course, it's possible that Columbus merely heard about such an event. He also knew about wood carvings and other objects of human origin that occasionally washed onto European shores. Though it wasn't necessarily clear that these had crossed the ocean, for Columbus, the mysterious flotsam—human and otherwise—was a series of clues adding up.

Around 1477, Columbus may have made it as far as Iceland, just a few decades after contact had been lost with the Viking colony in Greenland. Sailors in Iceland would have been aware of not only rocky islands across the Atlantic but an entire continent. Could this be the tip of Eurasia, as described by Marco Polo? If Columbus did make it to Iceland, this may well have been the pivotal event in his decision to sail west. He was certainly not alone in his conviction that Asia could be reached this way. In 1474, a Florentine astronomer named Toscanelli sent a letter and map to the king of Portugal. This map, which was soon shown to Columbus, seems to be the source of the miscalculation of the size of Earth. As we've seen, the Greek philosopher Eratosthenes had computed the circumference of the world to within 10 percent of its correct value. Toscanelli was either unaware of this calculation or didn't believe it.

Contrary to a view promulgated by Washington Irving's 1828 Columbus biography, almost no one at the time thought Earth was flat. The sphericity of Earth was the basis of medieval astronomy and would have been second nature to sailors who navigated by the stars and observed the horizon's curvature on a daily basis. Where did the error in distance arise? The Persian astronomer Alfraganus had calculated a degree of latitude (and longitude at the equator) to be fifty-six miles, but Toscanelli didn't realize that this referred to Arabic miles instead of Roman miles,* leading to an underestimate of 30 percent. Compounding the error, Toscanelli massively overestimated the width of Eurasia. Whereas Ptolemy had estimated that the continents spanned 180 degrees of longitude (half the world), Toscanelli based his calculations on a much larger

---

* A Roman mile (from "*mille passus*," a thousand paces) was originally 5,000 Roman feet (around 4,860 modern feet), whereas an Arabic mile is 6,584 feet.

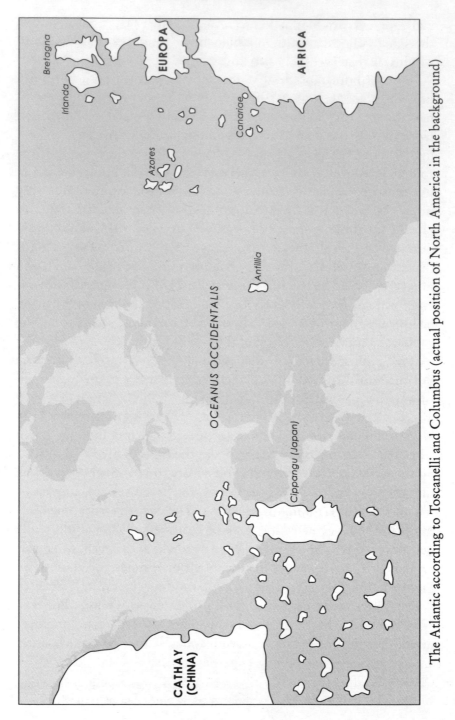

The Atlantic according to Toscanelli and Columbus (actual position of North America in the background)

225 degrees (two-thirds). In fact, even Ptolemy had overestimated the size of Eurasia (the actual figure is around 130 degrees). The result was that Toscanelli lost more than half the world.

Furthermore, Toscanelli guessed that Japan ("Cipangu," as described by Marco Polo) was larger and farther out, so it should have been easier to reach. He therefore estimated the distance from the Canary Islands to Japan to be a mere two thousand three hundred miles. The true figure is more than twelve thousand miles, spanning North America and the Pacific. Columbus was therefore working under the assumption that Asia was more than five times closer than it actually was. There was probably an element of wishful thinking in putting Asia right around the maximum distance a ship could hope to sail, but since Columbus knew that there were lands across the Atlantic, why *couldn't* these be parts of Asia? While most scholars correctly concluded that a westward voyage from Europe to Asia wasn't feasible, Columbus thought he knew better. It was a mistake that would change the world.

During the 1480s, Columbus made repeated attempts to secure backing for a westward voyage. In 1484, he asked King John II of Portugal for three ships to sail out into the Atlantic. The plan was summarily rejected, for the very good reason that the best geographers in Europe knew that even if a westward route were possible, Asia was much farther than Columbus assumed. Columbus next tried Spain in 1486, which had been unified only seven years prior by the marriage of Ferdinand II of Aragon and Isabella I of Castile. He presented his plan to Queen Isabella, but she rejected it for the same reason. Columbus tried Portugal again in 1488 after Bartolomeu Dias returned from the southern tip of Africa, but with a sea route to India at their fingertips, the Portuguese had no interest in far-fetched schemes. He tried Genoa, Venice, and even King Henry VII of England, all to no avail.

In fact, Columbus had piqued Queen Isabella's interest, but Spain was still embroiled in a struggle against the Moors. In 1489, to keep him from taking his ideas elsewhere, she gave Columbus an annual allowance and opened tentative negotiations. Finally, in January 1492, after the conquest of Granada, Ferdinand and Isabella summoned Columbus. According to the story, they initially turned him down one last time, but as he was leaving town in despair, they changed their minds and sent a royal guard to catch him. A deal was struck: Spain would pay for the voyage, and should it succeed, Columbus would be viceroy of any lands discovered, receive 10 percent of all revenues in perpetuity, and have the authority to nominate governors of the lands with royal approval. Why did the Spanish monarchs change their minds? Surely they must have considered the voyage an outlandish scheme, but there was at least a small chance that Columbus was right and everyone else was wrong. Having united their country, and eager now to gain a competitive edge over the rival Portuguese, who were on the cusp of reaching India, they must have judged Columbus's plan a fair gamble: likely to lead to nothing, but with just a tiny chance of a massive payoff.

Despite their commitment, Ferdinand and Isabella took many months to get organized. Far from pawning the crown jewels to pay for the voyage (as a popular story goes), the monarchs merely commanded the small seaport town of Palos de la Frontera to provide the ships and crews in repayment of a royal debt. Palos managed to scrounge up three barely seaworthy ships: two caravels, *Pinta* and *Niña*, and one carrack, *Santa María*. These were small vessels: at fifty feet long, *Pinta* and *Niña* were the length of a railway freight car, with less cargo capacity. *Santa María* was only 20 percent larger. These would be the size of pleasure yachts today—much smaller than many contemporary European ships and dwarfed by the massive Chinese treasure ships that Zheng He had sailed around the Indian Ocean sixty years earlier. Uncomfortable, cramped, and wet, the ships would have been tossed around in calm Atlantic waves, let alone a storm.

The small fleet left port on August 3, 1492, and first sailed to the Canary Islands to restock supplies and catch the prevailing winds.* Then, on September 6, they set out into the unknown. Columbus may have been wrong about the distance to Asia, but he was a skilled navigator. During the voyage, he was able to accu-

---

* Around the Canary Islands, the "easterlies" blow toward the west, while the "westerlies" blow back to Europe farther north, forming a giant circular conveyor in the North Atlantic.

rately track the ships' traveled distance, even though no one had yet invented a way to measure longitude. Theoretically, distance is just speed multiplied by sailing time. Later sailors could measure their speed by letting out a knotted rope into the sea and counting how long it took to unravel (hence speed is measured in "knots"), but this method wouldn't be invented for another century. Columbus had to judge his ship's speed by visually estimating how fast waves were passing, a method that not only requires tremendous attention and experience but also fails to correct for currents and wind patterns. Meanwhile, time was kept using sand hourglasses that had to be reset every thirty minutes by cabin boys who weren't notorious for adherence to duty.

By early October, Columbus and his men had sailed without sighting land for a month, and the crew was getting restless. They weren't so much worried about falling off the edge of Earth as worried that Columbus was wrong about the distance to Asia, and that they'd sail too far to make it back against unfavorable winds with their dwindling supplies. Passing through the Sargasso Sea,* the ships were slowed by sheets of floating seaweed that tangled against the hulls and glowed eerily at night. Finally, Columbus agreed that they'd turn back if land wasn't sighted within a week. On October 7, the crew spotted a welcome sight, when they recorded a shorebird flying to the west, presumably in the direction of land. Then, early in the morning of October 12, after five weeks at sea, a lookout on *Pinta* proclaimed the first sighting of land: "*Tierra!*"†

Later that day, the fleet anchored off an island they called San Salvador: an island of the Bahamas, but we aren't sure exactly which. Columbus waded ashore and claimed possession of the island in the names of Ferdinand and Isabella, and members of the crew signed as witnesses to the proclamation. Soon, curious islanders emerged, and the official Arabic translator (a con-

---

* The only sea on Earth not bounded by land, whose boundaries are defined by the swirling currents of the North Atlantic.

† According to some accounts, this lookout, Rodrigo de Triana, was denied a lifetime pension promised to the first sailor to sight land because Columbus later maintained that he himself had seen a light on the horizon a few hours earlier.

verted Jew) was summoned, but, to their great disappointment, he couldn't make out the native language. Nor did the inhabitants wear Chinese robes as expected; instead, they went practically naked. These were the first signs that the fleet may not in fact have reached Asia. On the other hand, perhaps this was an outlying island, so Columbus wasn't much dissuaded. The "Indians" were peaceful and friendly, but the Spanish noted many scars on their bodies and were able to deduce that people from other islands would occasionally come to capture them. These raiders were the Kalinago people, who'd migrated from South America into the Caribbean a few centuries earlier. The Kalinago word for "person" was "*karibna*," which served to name not only the Caribbean but also one who eats his fellow men, a "cannibal."*

The Spanish observed that the natives seemed happier than Europeans, living a simple existence with no knowledge of money, technology, or weapons. This was perhaps the first incarnation of the idea (some would say myth) of the "noble savage" later described by philosophers. Shortly after Columbus's voyage, the noble savage became a stock archetype in works of fiction as a way to reflectively criticize the ills of one's own society, right up to James Cameron's 2009 film *Avatar*. The French writer Michel Montaigne's 1580 essay "Des Cannibales" ("The Cannibals") was an early attempt to reverse the egocentric European belief in the superiority of Western culture. Based on descriptions and interviews of South American people, Montaigne described not an ignorant, barbarous culture but one emphasizing the virtuous skills needed to live in natural harmony. Cannibals they may be, Montaigne admitted, but European society was guilty of worse. While cannibals consume the flesh of the dead, Europeans torture and burn people alive for their religious beliefs.

Columbus noted that the gentle indigenous people would make ideal servants, so he kidnapped a few to present to Ferdinand and Isabella. But first, observing the natives' golden orna-

---

* The claim that the Kalinago were cannibals is based on little surviving evidence. It's possible that the story was fabricated by the Spanish to help justify enslaving them.

ments, he insisted that they guide him to the source of the gold. The natives indicated that it came from a large island to the south, ruled by a wealthy king. Delighted, Columbus assumed this to be Cipangu (Japan), described by Marco Polo as having cities lined with gold (though Marco was recounting from secondhand Chinese sources). Instead, the island turned out to be Hispaniola, where Columbus eventually did find some gold, but not nearly enough to satisfy his lofty expectations.

For nearly five months, Columbus explored the Caribbean. As the fleet encountered islands, Columbus sent scouting parties ashore, asking the locals about the great khan of China, inquiring as to the direction of his royal court. To these questions (communicated in gesture, we must assume), the Spanish received only puzzled looks—but they did find other interesting things. Scouts reported that the natives slept on beds made of cord hung between two trees, called "*hamacas*" (meaning "fishnets" in the local Taino language). Many a seasick sailor would owe thanks to these Caribbean people, as hammocks were soon introduced to all the sailing ships of Europe. In rough seas, hammocks remain in place while the ship rocks around the sleeper. The Spanish also found that locals carried small cylinders filled with dried leaves that they set afire at one end so that they glowed. People sucked on the other end, and, as it were, "drank" in the smoke. The natives called these cylinders "*tabacos*."

At Hispaniola, the expedition was warmly received by the native leader Guacanagaríx, one of five Taino chiefs on the island. The Spanish recorded that the Taino never fought among themselves but instead settled disputes by gathering a group of supporters for each side to compete in a soccer-style ball game. This may have been a misinterpretation, but it certainly is true that ball sports were popular in the New World, which absolutely fascinated the Spanish. Team sports had been popular during Roman times, but since then they'd virtually disappeared from Europe. While they were visiting Hispaniola, tragedy struck on Christmas Day 1492, as Columbus's largest ship, *Santa María*, ran aground on a reef. Since *Pinta* and *Niña* couldn't carry all of *Santa María*'s crew, Guacanagaríx gave permission for the Spanish to leave thirty-nine men behind in Hispaniola. After they hastily erected the settlement of La Navidad from *Santa María*'s wreckage, *Pinta* and *Niña* sailed for home.

The return voyage was a grueling affair, as the two tiny caravels were wracked by fierce winter storms. Sailing on *Niña*, Columbus was separated from *Pinta*, which went on ahead, and storm damage forced him to make emergency repairs in the Azores. There, the crew were imprisoned by the Portuguese governor on "suspicion of piracy," but Columbus was able to negotiate their release. Another storm forced *Niña* to seek refuge in Lisbon, where Columbus was interviewed by Bartolomeu Dias, whose voyage three years earlier had complicated Columbus's appeal for Portuguese support.* The Portuguese accused Columbus of violating the 1479 Treaty of Alcáçovas, which granted them rights to all future territories discovered in the Atlantic. Fortunately, they let the matter drop, and Columbus sailed on to Spain.

Columbus was received with great enthusiasm by Ferdinand and Isabella, who were eager to hear about the wonders of Asia. Although the voyage fell short of an alliance with the great khan, there seemed to be plenty to celebrate. Along with a few curiosities, such as hammocks and tobacco, Columbus had brought back native plants, birds, and even people as showcases of the Orient, plus at least a little gold. True, he hadn't found the magnificent golden palaces and marble cities that Marco Polo had described. A few questioning voices were raised, but not enough to dampen the general jubilation. Spain was now poised to beat the Portuguese to India, which surely lay just a short sail from the islands Columbus had discovered.

Columbus would lead three more voyages. The second aimed at settlement, sailing with seventeen ships and twelve hundred colonists in September 1493. Returning to Hispaniola, Columbus found La Navidad in ruins, its inhabitants killed in a struggle with locals. The fleet then sailed eastward along the coast of Hispaniola, establishing the settlement of La Isabela. However, even this location was

---

* Columbus probably also met the future world circumnavigator Ferdinand Magellan, who was a young ward of the court.

soon abandoned, its population moving in 1496 to the better-situated Santo Domingo (today the Dominican capital). Columbus continued to search the Caribbean for signs of Asia, trying to match what he saw with what Marco Polo had described, to no avail. Returning to Hispaniola, he governed the colony for a while but was a lousy administrator. With the settlement on the brink of rebellion, he skulked back to Spain in 1496 to appeal for additional support.

Columbus returned to the Spanish court, once again surrounded by the wonders of the Caribbean, but this time his reception was considerably cooler. After years of searching, he still had no evidence of having reached Asia. Yet, there were hopeful signs. On his third voyage, in 1498, Columbus found gold nuggets the size of hen's eggs in Hispaniola and reached the mouth of the Orinoco River, which disgorged enormous volumes of water that had to signify a continent. Here, Columbus made his crew solemnly swear that they'd reached Asia, as if believing might make it true. On his fourth voyage, he reached the Yucatán, where he finally received news that the splendors of Asia were at hand: the locals described magnificent inland cities with formidable palaces, ruled by nobles clad in gold. Surely this, at long last, must be the glorious Cathay described by Marco Polo.*

Meanwhile, Columbus's troubles worsened. In 1500, Ferdinand and Isabella dispatched a representative to investigate accusations of brutality and gross mismanagement in the Hispaniola colony. He reported that Columbus used torture and terror to maintain control, and enslaved the natives on a massive scale. Columbus was arrested and hauled back to Spain in chains to answer for his crimes. Eventually, he convinced the monarchs to set him free, but not before being stripped of his title and privileges. In the final years leading up to his death in 1506, Columbus became embittered, obsessed with being denied what he considered his rights. After his death, his heirs sued the Spanish Crown for the profits originally promised, but the monarchs considered these forfeit due to Columbus's crimes and miserable governorship. He died

---

* It was actually a reference to the Aztec capital of Tenochtitlán (Mexico City).

still believing he'd reached Asia, long after almost everyone else had guessed the truth about the new land.

Columbus's greatest legacy was that which he denied his entire life. Although he can't be said to have "discovered" the Americas in any realistic sense—they'd already been discovered by Europeans and were already filled with people—there's little doubt that he initiated a sustained exchange of people, animals, plants, and ideas that changed the world for both good and ill. Combined with the Portuguese voyages to Asia, the vast majority of Earth was now inextricably linked. American corn, beans, squash, and potatoes spread throughout the world, offering new ways to feed large populations. A more productive and diverse food supply all but ended the regular famines that had plagued Europe, and the continent's population exploded.

Columbus didn't invent slavery, but it was part of his legacy. Although disease was responsible for more deaths, the depopulation of Hispaniola amounted to deliberate genocide, with Spaniards brutalizing and murdering natives with impunity. Feudalism was revived in the New World as the *encomienda* system. Under *encomienda*, Spanish settlers were awarded a parcel of land, along with a native population to rule. *Encomienda* had developed during Roman times and was supposed to provide mutual benefit where lords provided protection in exchange for a share of the crop. In the New World, far from royal oversight, the system was rife with abuse. There was little incentive not to work the natives to death to enrich the landlord, who was rarely concerned with their long-term benefit. Queen Isabella had forbidden native enslavement, but *encomienda* was slavery in all but name.

Some Spaniards objected to *encomienda*, most notably Bartolomé de Las Casas. As a young man, Las Casas came to Hispaniola with his father in 1502 and participated in some raids against the native Taino. Around 1510, he became the first priest ordained in the New World and initially defended *encomienda*. Then, in 1513, he accompanied Spanish soldiers in the conquest of Cuba, where he witnessed the full brutality of several massacres. Through this, he became convinced of the grave injustice of Spanish rule, initiating a fifty-year campaign for universal human rights. He renounced his *encomienda* and preached that others should do the same. But, realizing that he'd have more impact in Spain, he returned there in 1515 to plead his case to the Spanish court.

Las Casas wrote pamphlets and books exposing horrible abuses by the Spanish in the New World and the utter destruction of native communities. He records that when he first arrived, there were hundreds of thousands of native inhabitants on Hispaniola, but most soon died from overwork, cruelty, famine, and disease. Instead of direct enslavement, Spaniards would sometimes tax the natives by imposing a quota of gold they'd have to produce every three months. Natives who came up with the gold were given a copper token to hang around their necks. Any native without a copper token was subject to mutilation, torture, or death. Meanwhile, European diseases drained the population, and in 1519 a smallpox epidemic wiped out many of the survivors. By 1548, only a few hundred remained on the entire island.

Although Bartolomé de Las Casas didn't end cruelty or slavery, his efforts achieved at least some results. He documented the atrocities in his writings *A Short Account of the Destruction of the Indies* (1542) and *General History of the Indies* (1552), both widely circulated in Europe. His royal lobbying eventually paid off, with the New Laws passed in 1542. These extended legal status and some protection to natives, although abuses were still common. Las Casas argued that the laws didn't go far enough, but the reforms were already so unpopular with settlers that they triggered riots. As slavery persisted and eventually shifted to the importation of Africans, Las Casas worked to open a dialogue on race and the ethics of colonialism. In the Valladolid debate of 1550, he resolutely argued that the conquest and subjugation of any race was unjustifiable.*

After a few decades in the New World, the Spanish became so obsessed with extracting gold and silver that they forgot about the spices that had motivated Columbus's Atlantic crossing. Instead, new generations of explorers extended Spanish dominion across

---

* In his early writings, Las Casas suggested importing Africans as an alternative to native labor. However, he eventually came to see all forms of slavery as abhorrent and campaigned against the use of Africans as well.

the Americas. Within a century, Spanish territory encompassed an area twenty-seven times larger than Spain itself. Almost none of this expansion was planned by Spain's rulers—rather, it was orchestrated by individual actors who led private military forces to explore and conquer newly discovered lands. These "conquistadors" were often minor nobles from rougher and poorer parts of Spain, denied opportunity for advancement back home. The New World offered the prospect of overturning a drab life in rural Europe, seizing unimaginable riches, and living on a vast estate built on free native labor. Some conquistadors vastly expanded Europe's geographic knowledge. Many met violent and ignominious ends. All left a trail of devastation in their wake.

In 1513, Vasco Núñez de Balboa became the first European to see the Pacific Ocean from the Americas, but he was beheaded a few years later by the governor of Panama. Also in 1513, Ponce de León landed in Florida to become the first European to set foot on the mainland north of Mexico since the Vikings, but he was killed by natives while trying to establish a settlement there in 1521. In 1539, Hernando de Soto trekked through the American South from Florida to Arkansas, becoming the first European to cross the Mississippi. After he died on the riverbank, more than half his expedition of seven hundred perished during a grueling twelve-hundred-mile march through Louisiana and Texas to safety in Spanish-held Mexico. At almost precisely the same time, Francisco Vázquez de Coronado led a similar expedition through Arizona, New Mexico, Texas, Oklahoma, and Kansas, approaching within a few hundred miles of de Soto's route from the opposite direction. He became the first European to see the Grand Canyon but never quite succeeded in finding the mythical Seven Cities of Gold he was looking for.

The most successful conquistadors led tiny bands to conquer powerful civilizations. Hernán Cortés was the first of these. Cortés was born on a rural estate in Extremadura (near Spain's border with Portugal) to a minor noble family. As a boy, he may have witnessed the procession that passed nearby after Columbus's return from the New World in 1493. In 1504, Cortés moved to Hispaniola to become an *encomienda* plantation owner, but he soon found the life

154

of a settler boring. In 1511, he joined the expedition to Cuba and, thanks to his military skills, was appointed to lead a 1519 invasion of Mexico. Soon after setting out, the governor revoked the appointment and recalled him to Cuba, but Cortés disobeyed orders and pressed on. When he reached the shores of Mexico, Cortés fought the Spanish troops who had been sent to arrest him (they joined him instead), burned his ships so no one could contemplate retreat, and marched off toward the capital of the Aztec Empire.*

Then, as the story goes, this band of a few hundred Spaniards, accompanied by fifteen horsemen and a few cannons, defeated an empire with a population in the millions. Of course, this story is nonsense. Despite the Spanish advantages of guns, horses, armor, and steel blades, if the Aztecs had wanted to destroy the Spaniards, they could easily have done so. Instead, Emperor Montezuma was curious about the newcomers and kept sending them gifts to placate them. This had the opposite effect of whetting their avaricious appetites. More important, as the Spanish marched on the Aztec capital, they found a native population willing to join them. The Aztecs had for centuries operated a tribute empire throughout Mexico, whose inhabitants had to pay heavy taxes to Tenochtitlán in the form of both goods and people. Human sacrifice had been an integral part of earlier civilizations such as the Maya, who believed it was necessary to sustain the gods of the universe, but with the Aztecs it was practiced on an unprecedented scale.

The Spanish estimated that as many as twenty thousand people were sacrificed in Mexico every year, although this figure could have been exaggerated. It's sometimes claimed that sacrificial victims understood that they were performing a service, and some went willingly, but this seems unlikely to have been true for most. The Spaniards described Aztec sacrifice rituals as barbaric, and indeed they were. However, European executions were also public spectacles of ritualized religious rites involving an appeal to the

---

* Technically, there was no such thing as an "Aztec Empire." The people who ruled Tenochtitlán called themselves "Mēxihcah" (the origin of "Mexico"), and the empire was an alliance of three political entities. The explorer Alexander von Humboldt (whom we'll meet later) popularized the term "Aztec" in 1810 to describe the people linked to this "Triple Alliance."

heavens for salvation, culminating in violent death. Twenty thousand victims per year may seem high, but if we assume a Mexican population of twenty million, this represents a mere 0.1 percent of the population, which is roughly equivalent to the execution rate in Europe at the time. Regardless, it's clear that the Aztecs had lots of enemies, and many native Mexicans sided with Cortés against their former oppressors. When the Spanish marched on Tenochtitlán, they were accompanied by thousands of native allies.

However, it still would have been easy for the Aztecs to defeat the smaller Spanish and native force. Instead, Montezuma invited the Spaniards to the capital so he could learn more about them, whereupon they took him prisoner. Thereafter, Cortés used Montezuma as a puppet, compelling him to surrender his treasure and disarm his army. Even then, it was a struggle for the Spanish, because the priests organized the Aztec armies to expel the Spanish. Cortés and his men barely escaped alive, after a running battle through the streets and canals of Tenochtitlán. The Spanish rear guard was massacred. Eventually, with reinforcements from Cuba, more native help, a smallpox epidemic that decimated the Aztec population, and the months-long starvation of Tenochtitlán, the Spanish finally managed to conquer the city. Even with the advantages of superior technology, ruthless cunning, thousands of native allies, and deadly disease, it was a contest the Spanish nearly lost. In the end, the Aztecs were largely defeated by their own ill treatment of the native people of Mexico, who sided with Cortés when the opportunity came.

The other conquistador to topple a fabulously wealthy civilization came from even humbler beginnings. In 1532, Francisco Pizarro, an illegitimate and illiterate swine herder from a poor Spanish family, led 182 men to conquer the Inca Empire of 12 million inhabitants. This seems impossible, and indeed it was. The Incas had a powerful army of tens of thousands, but it had just been ravaged by newly arrived European diseases and a civil war. Even so, the Spanish only succeeded because instead of taking them seriously, the Incan emperor, Atahualpa (who'd won the civil war against his brother Huáscar), treated the Spanish as mere curiosities. When he met them at Cajamarca, Atahualpa was accompanied by an army of fifty thousand, but he chose to approach the

Spanish with a small number of ceremonial guards. Pizarro's men ambushed the Incas, hacking them to pieces with steel swords. Taking Atahualpa hostage, they neutralized the Inca chain of command precisely as Cortés had done with the Aztecs.*

Yet even with Atahualpa as a puppet, conquering the Inca Empire wasn't easy. After forcing Atahualpa to fill one room with gold and two with silver, the Spanish murdered him and marched on the capital of Cuzco. By this time, they were accompanied by more soldiers (at least five hundred) and also some native allies. Nevertheless, although the Spanish quickly occupied Cuzco, the remnants of the Inca Empire organized an effective resistance for forty years. For decades, Inca bands would attack Spanish convoys along the mountain roads of Peru, rolling boulders down onto the Spanish and pelting them with rocks thrown by slings. Even in direct combat, the Incas were relatively effective against the Spanish, using a wide array of weapons, such as arrows, javelins, and copper maces and axes. Incan woven cloth armor was so light, and yet strong, that it proved perfect for the mountainous terrain of the Andes. Far from a quick and easy victory, the conquest of Peru was a struggle that stretched out for decades.

For the native inhabitants of the Americas, the arrival of Europeans was utterly devastating. Up to 90 percent of the population was wiped out in history's largest human catastrophe. Meanwhile, Spain carved out a vast empire based on the extraction of gold and silver. These riches would, in the short term, fill the coffers of Spanish kings and finance European wars from Italy to the Netherlands as Spain flexed its imperial muscles. Yet, the precious metals pouring in from the Americas would also drive an inflationary crisis that would, in the long run, sap the economy and lead to a collapse of Spanish power. Meanwhile, the Americas were forever thrust into a growing network that would soon span the entire world.

---

* Indeed, Pizarro knew about the conquest of Mexico and deliberately applied similar means. How differently it would have turned out if Atahualpa had known about the Aztec-Spanish encounter!

12 | AROUND THE WORLD

Columbus went to his grave believing that he'd reached Asia, but by the early 1500s, it was becoming clear to everyone else that the lands he discovered were in fact new continents. In 1497, another Genoese navigator named Giovanni Caboto sailed as "John Cabot" for Henry VII of England to look for a northern route to Asia. Sailing around Newfoundland and along the coasts of Labrador, Nova Scotia, and Maine, Cabot reported a thickly wooded land stretching for hundreds of miles. A few years later, in 1500, the Portuguese navigator Pedro Álvares Cabral accidentally discovered Brazil as he was sailing far out from the coast of Africa on a voyage to India. Added together, these discoveries were starting to suggest something more than a few scattered islands.

In a series of four voyages between 1497 and 1504, the Italian navigator Amerigo Vespucci sailed along the coastline from the Caribbean to Brazil. In 1500, his second expedition discovered the mouth of the Amazon. This giant river, disgorging a fifth of all fresh water flowing in all rivers on Earth, dilutes the ocean's salinity for miles. Even to Renaissance Europeans, this signaled an enormous landmass. As Vespucci traced the coastline, he became convinced that these lands were in fact continents, popularizing this idea with his letter "Mundus Novis" ("New World"), which was printed and distributed throughout Europe. Soon, the German cartographer Martin Waldseemüller labeled the new lands with the Latin version of Vespucci's first name in his 1507 atlas *Universalis Cosmographia*. America was born, named for a little-known Italian who had the courage to venture into the unknown.

Once it became clear that the Americas were continents, the next question was whether there was a way to get around them. It was thought that there must lie an ocean on the other side of the landmass, and the Portuguese who'd reached the Orient by sailing around Africa had reported what seemed like a major ocean to the east of the Spice Islands. The year 1513 brought the sensational news that the anticipated ocean had at last been discovered, when Vasco Núñez de Balboa trekked through the jungles of Panama to sight the Pacific for the first time. If only there were a passage or channel through the Americas, might Asia still be reached by sailing west?*

An ambitious Portuguese naval officer thought he knew the answer. Fernão de Magalhães (rendered in English as Ferdinand Magellan) devised a scheme to sail south along the South American coast until he found a passage. From there, he thought, it should be a short sail across the Pacific to the Spice Islands. We're not sure what Magellan thought he knew, but he might have been influenced by a 1516 Spanish expedition that had sailed as far as the La Plata River estuary—a 150-mile gap in the continent that could easily be mistaken for a passage. Magellan took his plan to the Portuguese king, but the king wasn't interested in sailing west since Portugal had already reached the Spice Islands. So Magellan took his plan to the rival Spanish, who were keenly interested in breaking the Portuguese spice monopoly.

In his misplaced optimism, Magellan mirrored Columbus. Like the Genoese navigator, Magellan believed that he alone possessed secret geographical knowledge contrary to all expert opinion, and was determined to shop around until he found support for his scheme. Like Columbus, Magellan continued to massively underestimate the size of the world. In his imagination, the world's larg-

---

* Panama was narrow enough to contemplate digging a canal, but such thinking was forbidden by the Spanish throne as being contrary to divine will—if the creator had wanted a passage, he would have made one! Instead, the Spanish transported materials overland through Panama for the entire colonial period.

est ocean shrank to a few days' sailing. Thus, two of the greatest explorations in history were grounded in profound ignorance. In geography, as in science, sometimes the greatest discoveries arise from the greatest errors.

As had occurred with Columbus, the Spanish approved the voyage but weren't interested in spending a lot of money. Magellan was given a small fleet of five ships optimistically named the "Armada de Molucca," but the ships were in such rough shape that he wondered if they'd been scavenged from naval graveyards. The crew wasn't much better: a hodgepodge of sailors from many nations was assembled, some scrounged from Spanish dungeons. Magellan never told them the destination. Had the crew known that he was loading provisions for history's longest voyage, they might have preferred to stay in prison. The fleet set out on August 10, 1519, and proceeded down the coast of South America. Stopping in Rio de Janeiro to make repairs, they enjoyed the hospitality of the locals. Technically, this was an illegal stop since the region had been awarded to Portugal, but it hadn't been settled by the Portuguese yet. After a month in Rio, they set out on what Magellan assumed to be the final leg of the journey: a dash through the passage, and then a quick sail across the Pacific.

As they sailed down the coast, it became more and more evident that the passage Magellan was counting on simply didn't exist. With each inlet and bay, the fleet wasted valuable time scouting to discern whether it was the passage they sought. This created a problem for Magellan, as he couldn't pass up opportunities to look for the passage but also couldn't reveal to his crew that he had no idea where to look. The biggest disappointment was the La Plata estuary, which turned out to be the opening of a massive river mouth—which they would have realized immediately if they'd simply tasted the fresh water, even miles out at sea. After months of fruitless searching, the crew had finally had enough. On April 1, 1520, while overwintering in Argentina, three Spanish captains mutinied and tried to seize control of the fleet.

Magellan sprang into action, capturing one of the rebel ships to shift the odds to three ships for Magellan versus two for the mutineers. After a brief standoff, the rebels surrendered to Magellan, whose revenge was swift. He ordered one of the rebel captains executed, the sword wielded by the captain's servant in exchange for a pardon. The other rebel captain was marooned, never to be seen again. It was a tough situation: The Spanish captains saw their fleet being led to its doom by a reckless foreigner. Was their duty to follow orders or to intervene?

By this point Magellan was obsessed, committed to the mission regardless of the cost. Soon after the mutiny, tragedy struck when the smallest ship, *Santiago*, sank in a storm. Miraculously, the crew survived, and two sailors straggled overland to the winter camp to get help. These men weren't the only people to wander into camp: a few natives also approached to satisfy their curiosity and trade with the visitors. Large in stature, the natives seemed like giants to the Spanish, who called them "Patagón," for "big feet," christening the land "Patagonia." (Modern estimates put their height at six and a half feet—not giants, but very tall for the standards of the time.)

The voyage resumed on August 24, more than a year after leaving Spain. At long last, a month and a half later, the expedition finally stumbled upon a seawater channel that might be the long-sought passage. With supplies dwindling, the other captains urged Magellan to turn back, mission accomplished, and have a follow-up expedition follow the strait to Asia. But Magellan was determined to press on. The strait separating the southern tip of South America from Tierra del Fuego* is a long and narrow channel, with rough seas and many bends and turns. For a month and a half, the fleet slowly felt its way through the passage at great peril. As Magellan was scouting ahead for the best route, the captain of *San Antonio*, Estêvão Gómez, decided he'd had enough and turned for home.† At the end of November, Magellan's three remaining ships finally entered the Pacific. Despite dwindling

---

* Named "Land of Fire" for the natives who kept fires burning all the time, as noted by Magellan's crew.

† Upon reaching Spain in May 1521, Gómez and his crew were imprisoned, but were freed when the last surviving ship of the expedition finally straggled into port more than a year later. Gómez then led his own expedition to try to reach the Spice Islands via a northwest passage but only managed to explore Maine and the Hudson River.

supplies, Magellan decided to press on, anticipating a quick sprint across the Pacific to the Spice Islands.

The voyage was the worst kind of hell—the longest journey until that time out of sight of land. The winds were favorable and the fleet made good progress, but covering more than a third of Earth's surface takes time. For more than four months, the crew sailed across the Pacific, somehow missing every island along the way. The water was putrid and almost undrinkable. The few remaining biscuits were moldy and weevil infested. The dried meat was rotten and glowing with phosphorescent bacteria. The crew was reduced to eating boiled strips of leather and trading captured rats as delicacies. Starvation and scurvy had cut the crew in half by the time they finally sighted land on March 6, 1521. This was the island of Guam, but its inhabitants proved unfriendly. The Italian diarist of the voyage, Antonio Pigafetta, recorded that they swarmed the ships and tried to rob the weakened crew of anything that wasn't fastened down.

The fleet sailed on to the next island, which they reached a few weeks later. Here, at last, the crew found sustenance, as friendly islanders paddled out to greet them with boatloads of exotic fruits. Magellan met with the local chief, and to his surprise, his servant Enrique could for the first time communicate with the natives, albeit with difficulty. They decided they must be close to Enrique's home in Malaya, and Magellan at first assumed that they'd reached the Spice Islands.* However, they'd sailed too far north, and these islands were in fact the Philippines (later named after Spain's King Philip II). In exchange for hospitality, Magellan agreed to intervene in local politics by fighting a rival tribe on the island of Mactan. A force of forty-nine Europeans led by Magellan waded ashore at Mactan in battle armor to face off against five hundred native warriors. Although the American conquistadors had prevailed against far greater odds, here it was not to be. The natives overwhelmed Magellan's force with spears, bows, and darts. Recognizing the leader, they brought down Magellan in full plate armor, still fighting ferociously.

---

* In fact, Enrique was the first human to circumnavigate the globe, having originally come from Southeast Asia (he'd been captured by Magellan at Malacca in 1511).

Deprived of its leader and short on men, the fleet burned one ship to distribute the crew among the two remaining vessels. Finally, after another desperate voyage groping about aimlessly in unfamiliar waters, the skeleton fleet managed to reach Brunei on the island of Borneo. Here, the diarist Pigafetta described a wealthy kingdom with a modern military force including many cannons and trained elephants. They were now entering the known realm that the Portuguese had already reached by sailing east around Africa. Thankfully, the sultan of Brunei proved friendly and supplied the ships with provisions, possibly thinking the Spanish useful allies against Portugal. Setting sail again, finally, at long last, more than two years after departing Spain, the two remaining ships dropped anchor off Tidore, one of the Spice Islands. Fewer than half the men had managed to make it this far, but now the survivors enjoyed rest in a tropical paradise.*

Not wanting to risk another epic Pacific crossing, the crews of the two remaining ships decided to try to reach Spain by sailing west. This decision was effectively an admission that the Portuguese route across the Indian Ocean was superior, although it would prove all the more dangerous for rival Spanish ships, which would have to sail through enemy territory. The ships were loaded with a valuable cargo of spices and set sail in January 1522. However, soon *Trinidad* began to take on water. The smaller *Victoria* was not large enough to carry all the men, so they split up. While *Victoria* went on alone, some stayed behind to try to repair *Trinidad*. Months later, *Trinidad* and her crew were captured by the Portuguese, who imprisoned them for years. Only a few ever made it home. Nevertheless, the captives may have been the lucky ones.

Victoria's voyage back to Spain was even worse than the marathon journey across the Pacific. For more than six months, the

---

* The Spice Islands were still ruled by native chiefs, but some Europeans had moved there as civil administrators. Foremost among them was Francisco Serrão, one of Magellan's friends from Portuguese service and the brother of Magellan's most loyal ship captain, João Serrão. The brothers were hoping to reunite when the fleet reached the Spice Islands, but both died around the same time in 1521, a few months before this could happen.

captain, Juan Sebastián Elcano, kept the ship far from shore, fearing capture by the Portuguese. By the time *Victoria* rounded the Cape of Good Hope at the southern tip of Africa in May, the food was almost gone and the sick were begging to be dropped off at the nearest Portuguese outpost. Finally, *Victoria* made a brief stop at Cape Verde (350 miles off of Africa's western bulge) on July 9, after 187 days at sea. When *Victoria* had set out from Tidore, she had carried more than 50 men. Now, from an initial crew of 270 in the expedition, only 18 starving invalids managed to straggle home aboard *Victoria*.* For the first time, a ship had circumnavigated the globe and proven the sphericity of Earth. This was surely one of the greatest exploratory feats of all time and one of the greatest tests of human endurance.† Yet, Magellan's strait was too remote to be useful, and the Spice Islands were thousands of miles beyond. In the end, the expedition proved the opposite of what it sought: there wasn't a practical route to Asia through the Americas after all.

The greatest legacy of Magellan's expedition was forging the final link in the chain binding the people of Earth. In the following decades, Spanish fleets would cross the Pacific to colonize the Philippines. The Manila outpost attracted not only Europeans but Chinese, Japanese, Indians, and Vietnamese. Galleons sailed between Manila and Acapulco for the next 250 years, exchanging American silver for Chinese porcelain, lacquerware, and silk. Chinese products were transported overland to Mexico City and thence on to Europe. These routes carried not only goods but also people. In 1613, a samurai warrior became the first Japanese citizen to visit Mexico, and then to travel across the Atlantic to Rome. Chinese immigrants arrived in the New World, intermixing with Europeans and Native Americans. It was the start of the global interchange that pervades our world. It was the beginning of modernity.

---

* Four more eventually managed to make it back from the *Trinidad*, and around fifty had turned back on the *San Antonio*.

† Upon their return, the crew noticed that they seemed to be missing a day. Eventually, it was realized that losing a day was an inevitable product of sailing around the world.

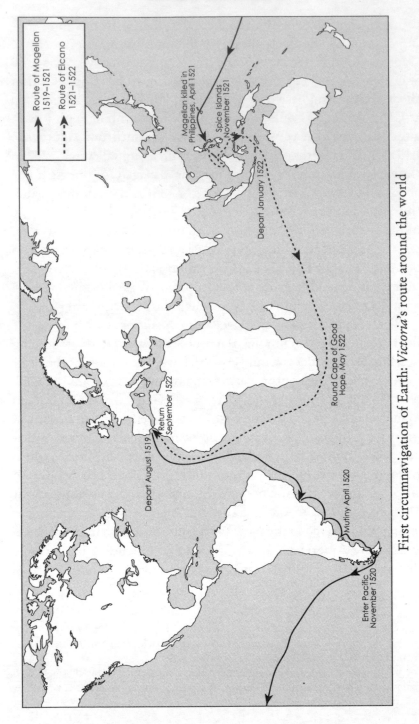

Legend:
Route of Magellan
1519–1521
Route of Elcano
1521–1522

Magellan killed in
Philippines, April 1521

Spice Islands
November 1521

Depart January 1522

Round Cape of Good
Hope, May 1522

Depart August 1519

Return
September 1522

Mutiny April 1520

Enter Pacific
November 1520

First circumnavigation of Earth: *Victoria's* route around the world

PART III

MODERNITY

13 | EMPIRES OF TRADE

Standing at its height in 1400, China was the leading power of the world. By 1800, it hadn't changed much. Cut off from the world since abandoning exploration, China stagnated, frozen in time. Its decline relative to Europe is perhaps best illustrated by the British diplomatic mission of 1793. In that year, George Macartney traveled to Beijing as Britain's first envoy to China. One of the goals was to demonstrate the latest advancements of European science and technology in the hope that China would finally decide to open its markets to European goods. The expedition brought eighteenth-century technological marvels as gifts to the emperor, including telescopes, firearms, textiles, and mechanical clocks. The emperor already had a collection of clocks, though none so sophisticated as the new British designs. Nevertheless, he wasn't especially impressed and simply added them to the imperial vault, where they collected dust. China had become inward looking, with no interest in progress.*

In 1400, Europe accounted for a mere 10 percent of the world's territory and 15 percent of its population. By 1900, it controlled 60 percent of the world's territory and population, and 75 percent of its economic output. Why did Europe take a different path than

---

* China drafted a letter to King George III declaring, "Our Celestial Empire possesses all things in prolific abundance and lacks no product within its borders. There is therefore no need to import the manufactures of outside barbarians."

China? Part of the answer is geography. While China encompasses a wide plain with two giant rivers flowing in the same direction, Europe is a jagged peninsula with an extremely long coastline beset by many bays, inlets, and channels. Europe's rivers flow outward in all directions, feeding a dozen seas. Its waterways, mountain ranges, and forests naturally divide the continent into segregated parcels, encouraging trade. While China was a vast united empire, Europe was a patchwork of squabbling states. This meant constant competition in Europe, which drove technological and economic progress. While China was the source of many inventions, European states perfected them in hopes of gaining a competitive advantage over their adversaries.

With a centralized imperial bureaucracy, there was little incentive for China to innovate, so it stagnated for centuries. Meanwhile, Europe was locked in a perpetual struggle for supremacy. Britain, for example, was at war with someone in the world a staggering 94 percent of the time between 1337 and 1914—especially surprising considering this period includes the ninety-nine supposedly peaceful years of the "Pax Britannica" between 1815 and 1914. Although not all of these were major conflicts (the Anglo-Zanzibar War lasted thirty-eight minutes), constant struggle doubtless encouraged technical innovation, not only in the military but also in economic and scientific spheres. Even when not at war, Europe was usually preparing for it. Industrial, transport, and communication systems were planned with military considerations in mind. The Russian railway network was designed to transport troops to face Germany and Austria-Hungary in the west, or all the way to its borders with Japan and China in the Far East.

Competition between European states resulted not only in the rapid perfection of technologies but also in their rapid spread. The movable-type printing press was invented in China a thousand years ago, but its adoption there was slow. This is partly due to technical differences between Chinese and Latin scripts: while most European languages have only twenty-six letters, there are over fifty thousand unique Chinese characters. That's a lot of distinct blocks to manipulate. Books in China could still be printed by the page, and indeed they were, but in nothing like the volume of Europe. Within a few centuries, a dozen government print shops had produced perhaps ten thousand Chinese books. By contrast, adoption of the printing press was explosive in Europe, as each state scrambled to outdo the competition with its own mass-produced pamphlets and books. Printed books were the fifteenth-century version of the Internet, with almost as rapid a spread. In 1465, when Gutenberg printed his eponymous Bible, there were only three print shops in all of Europe. Just thirty-five years later, a thousand printing presses across two hundred fifty cities had collectively churned out over ten million books.

European literacy rates doubled in the second half of the fifteenth century, to around 10 percent by 1500. By 1600, the spread of books meant that literacy rates had shot up to 50 percent in Northern Europe, and continued to rise to around 90 percent by 1900. With this new means of self-empowerment, highly motivated individuals could, for the first time, educate themselves. Many scientists got their start by simply reading about the latest developments in widely circulated books and journals. In the 1700s, for example, the clockmaker John Harrison rose from obscurity to solve history's greatest navigational challenge. By developing a means of computing longitude for any point on Earth, he earned prize money equivalent to $5 million today.

The adoption of printing was slow in the rest of the world. Printing remained prohibited in the Ottoman Empire until 1729 and, initially, even carried the death penalty. In India, printed books were presented to Emperor Akbar in 1580, but apparently without arousing curiosity. There may have been some difficulty printing Indian scripts instead of a twenty-six-character Latin alphabet, but this can't be the only explanation. With enough incentive, printing would surely have been common in both India and China, as it is today.

Technologies were slow to evolve in monolithic empires because innovations were perceived as threats to those in power. Change comes from a desire to upset the status quo, not maintain it. The elites of eastern empires had little incentive to encourage innovation, whereas the patchwork of European states struggling for supremacy drove competition. Competition meant opportunity. If ideas weren't appreciated in one realm, there were always neighbors to try. Columbus petitioned at least four monarchs before securing backing for his voyage west. Magellan sailed for Spain because Portugal wasn't interested in upsetting its monopoly. In China, a single imperial rejection would have ended it all.

We sometimes imagine that Europe became more powerful than the rest of the world and then set out to explore and conquer it. This is telling the story from the perspective of hindsight. The Europeans who first set out during the Age of Exploration didn't yet enjoy significant superiority from a technological or economic perspective. In the Americas, Europeans may have possessed the advantages of guns, germs, and steel (as Jared Diamond brilliantly recounts), but this certainly wasn't true in Asia, where Europeans were still arguably inferior in these categories. Our view of the relationship between exploration and technological development is largely a reversal of cause and effect. It wasn't that Europe became more advanced and then set out to discover the world. Rather, it was by exploring the world and positioning itself to dominate trade that Europe gained a relative economic and technological advantage.

Whereas China shut down overseas trade in favor of stability, European nations went out into the world to encounter new ideas, products, and technologies, and in competing with each other, selectively adopted the best of these. Through their voyages and conquests, Europeans became arbiters of international exchange, acting as intermediaries even between countries that had previously traded with each other, such as India and China. And with Europeans controlling 60 percent of the world's population and 75 percent of its economic output by 1900, they were often able to tax the initial producers and final consumers as well.

It's no coincidence that the Age of Exploration accompanied the scientific revolution. Columbus and Magellan were testing hypotheses about the geography of Earth. The Portuguese were similarly testing a hypothesis about the possibility of reaching India by sailing around Africa. Exploration and science went hand in hand: the same drive that motivated overseas voyages stimulated discovery at home. As explorers traveled the globe and discovered new lands, new people, new plants, and new animals, they expanded our understanding of what was possible. The world became a laboratory for experiments on science, economics, social norms, and forms of government. At the same time, discoveries undermined traditional authorities. There were entire civilizations in Asia and the Americas never mentioned in religious texts. How reliable could such texts really be?

European competition meant that once Spain and Portugal established their overseas empires, rivals began to threaten them. Britain and France would eventually predominate, but for more than a century starting in the late 1500s, the world's economic and naval powerhouse was the tiny Dutch Republic.* The Dutch exemplified the best of the Renaissance competitive spirit. While most countries were stratified with the nobility and clergy on top, Holland supported a large middle class. Business, law, the arts, and all spheres of science flourished. Universities became pathways to public office. Dutch society was the most tolerant in Europe, as Holland became a gathering place for intellectuals fleeing religious and political persecution. Jews expelled from Spain were welcomed in the Netherlands. Scientific and philosophical works that were banned abroad were printed with zeal in Holland and often secretly exported. The great French philosopher and mathematician René Descartes actually lived in Holland and only visited his native country on business.

During the Dutch Golden Age, art and science flourished like never before. Until this point, paintings had mostly featured stuffy religious icons and biblical scenes. But Dutch masters such as Rembrandt and Vermeer rejected such narrowness, showcasing scenes of daily life that were based on meticulous experiments with geometry and light. Christiaan Huygens built on Galileo's pioneering astronomical work with telescopes, explained Saturn's rings, and invented pendulum clocks so that accurate time could be kept for the first time. Scaling down to the world of the

---

* Formed when seven rebellious Dutch provinces overthrew Spanish rule in 1568; its struggle for independence lasted until 1648. Spain retained the southern provinces, which remained mostly Catholic and became Belgium.

very small, Anton van Leeuwenhoek invented the microscope and observed that life is composed of cells. The universe was becoming more fascinating than anyone had imagined, with discoveries ranging from giant stars and planets down to the incredible diversity of life in the tiniest drop of pond water.

The Dutch simultaneously revolutionized economics, building the first stock exchange and first central bank, and inventing concepts such as insurance and retirement funds to pool assets and distribute risk. In 1602, they established the first publicly traded joint stock corporation, the Dutch East India Company. Resources were combined to fund expeditions, and spices were imported in bulk to be sold at huge profits. The Dutch had long been merchants, with access to river routes into the German hinterland, but now they became the leading naval power of their age. One and a half million Dutch, constituting a mere 0.25 percent of Earth's population, were poised to control most of the world's overseas trade from a nation half the size of Maine.

As the Dutch gained power, Spain and Portugal declined, and much of this was also due to economics. Prices rose precipitously across Europe throughout the 1500s, and at first no one could explain why. Common wisdom held that money was extracted from the ground, so Spain could become rich by looting precious metals from the Americas. For a while this worked, but instead of investing in their economy, the Spanish spent the money on religious wars and, also, on manufactured goods from Northern Europe. The net result was that, over time, Northern Europe ended up with both the manufacturing capacity *and* the money. Meanwhile, prices kept rising, as New World metals tripled the volume of gold and silver in circulation, so that far more money was chasing the same goods.* As Spain was forced to declare bankruptcy seven times between 1557 and 1653, some Spaniards wondered if their American empire was more a liability than an asset.

The shipyards at Amsterdam and Rotterdam turned out ves-

---

* It seems remarkable, but until the 1500s, no one had a concept of inflation. An early pioneer of the quantity theory of money was Copernicus (who famously reintroduced the Sun-centered solar system). In 1568, the French philosopher Jean Bodin finally attributed price increases to a surfeit of gold and silver.

sels in record time. By 1650, over sixteen thousand Dutch ships plied the world's oceans—by far the largest fleet in the world. The Dutch East India Company, acting almost as an independent state, established colonies and trading posts across the globe. In 1621, the company ejected Portugal from the Spice Islands and asserted direct control, slaughtering many of the native inhabitants. In 1641, the Dutch conquered the Portuguese fortress of Malacca, opening trade with China and Japan.* They established a resupply station at the Cape of Good Hope, where their Afrikaans-speaking descendants form the majority of white South Africa today. Between 1642 and 1644, Abel Tasman explored the South Pacific as far as Fiji and New Zealand, and circumnavigated Australia.

Perhaps the most famous Dutch settlement was New Amsterdam, founded in 1624 to trade with Native Americans along the Hudson River. Though the British occupied the colony in 1664, renaming it after the Duke of York, local names such as Brooklyn, Harlem, and the Bronx betray the city's Dutch heritage.† The legacy of New Amsterdam extends far beyond what you might expect for a colony lost more than three centuries ago and lasting a mere forty years. Around five million Americans trace their Dutch ancestry to its original settlers, and around 1 percent of the words in American English come from Dutch, including "booze," "boss," "coleslaw," "cookie," "cruise," "decoy," "dope," "kink," "spooky," and even "Yankee." Perhaps the most iconic appropriation of Dutch culture is Santa Claus, whose original incarnation, Sinterklaas, was a potbellied Dutch sailor—a lampoon of New York Dutch culture.

Soon the British and French joined the overseas scramble and dispatched pirates to harass Portuguese and Spanish shipping. Most famous was Francis Drake, who circumnavigated the world from 1577 to 1580, attacking Spanish ships and harbors along the

---

* The Dutch were the only Europeans permitted to trade in Japan during its two-hundred-year national isolation (1639–1853). When the Japanese broke their isolation and sent emissaries to Europe, the only European language they spoke was Dutch—which, to their surprise, wasn't all that useful.

† Wall Street ("Waalstraat" in the original Dutch) was named for an actual wall built to protect the city from Native Americans, pirates, and the British.

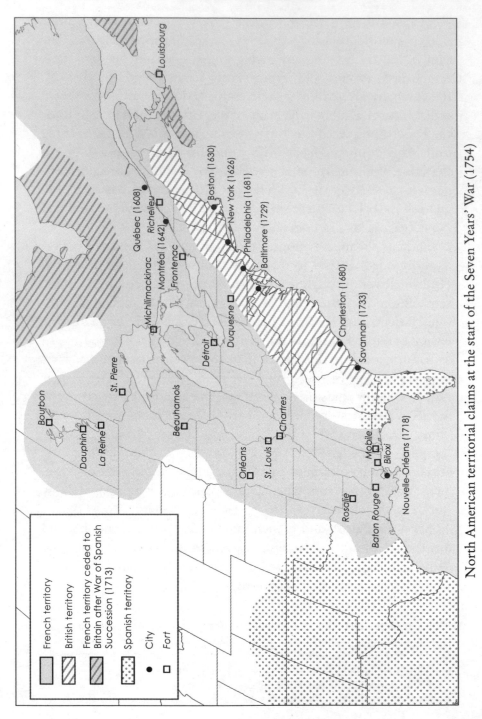

North American territorial claims at the start of the Seven Years' War (1754)

Québec (1608)
Richelieu
Montréal (1642)
Frontenac
Michillimackinac
Boston (1630)
New York (1626)
Philadelphia (1681)
Baltimore (1729)
Charleston (1680)
Savannah (1733)
Détroit
Duquesne
St. Pierre
Bourbon
Dauphin
La Reine
Beauharnois
Orléans
St. Louis
Chartres
Mobile
Biloxi
Nouvelle-Orléans (1718)
Rosalie
Baton Rouge
Louisbourg

French territory
British territory
French territory ceded to Britain after War of Spanish Succession (1713)
Spanish territory
City
Fort

way. A lesser-known Dutch captain, Olivier van Noort, repeated this feat two decades later. Like the Dutch, the British and French established joint stock corporations. While the British East India Company competed with the Dutch in Asia, the Mississippi Company attempted to settle French territory in North America; it collapsed when it turned out that no one wanted to move to swamp-infested Louisiana. In a rush to attract colonists, French authorities released inmates from Paris jails under the condition that they marry prostitutes and settle in New Orleans. The newly-weds were led to the port of embarkation chained together.

The vast center of the United States was once French territory, stretching from Canada through the Great Lakes down the Ohio and Mississippi valleys. Meanwhile, the British concentrated their colonies along the Eastern Seaboard. With fewer settlers spread across vastly more territory, it was perhaps inevitable that the French would lose in the end. The Seven Years' War (1756–1763) eventually decided the contest in favor of Britain.* Fought by European empires that were now global, this conflict was truly the first world war, waged on almost every continent and ocean. Such was the reach of Europe by the mid-1700s that a war in Europe meant conflict in not only the Americas but India, Africa, and the Far East as well. Britain won the contest, becoming master of the oceans, India, and North America, only to lose most of the latter twenty years later when its rebellious American colonies gained their independence.

The global reach of Europe spawned economic and technological development but came with negative consequences, including colonialism and slavery. As early as 1501, Spanish settlers began importing Africans to the Americas as native populations succumbed to disease. Soon, a "triangular trade" developed between Europe, Africa, and the Americas. Guns, ammunition, and manufactured goods were shipped from Europe to Africa to buy slaves. Slaves were in turn transported to the Americas to produce sugar, tobacco, rum, and cotton for export to Europe. Europeans didn't

---

* A better name would be "Nine Years' War" since it began two years earlier in America than in Europe (1754).

invent slavery—it had existed in most civilizations throughout history—but they did make it vastly worse by forcibly transporting people across the Atlantic and working them to death in horrific conditions. In the Americas, chattel slavery exploited Africans as property, no longer people but objects to be legally bought, sold, punished, or killed at will.

Africa was devastated by the slave trade. Over the course of three centuries, Europeans flooded Central and West Africa with twenty-five million guns. This influx of weaponry shattered the long-standing stability of African kingdoms, many of which had once rivaled European states in prosperity. Africans faced a stark choice: either participate in the "guns for slaves" trade, or risk falling victim to neighboring kingdoms that did. The damage to Africa went beyond mass abduction: it's been estimated that for every slave captured, another African was killed in violence fueled by the trade.

At times, slaves managed to seize power to direct their own destiny. Doing so was difficult, as slaveholders did everything they could to prevent rebellion: corporal punishment, enforced illiteracy, and as much segregation as possible. Nevertheless, slave revolts were common. Most famous in the United States was the 1831 Nat Turner Rebellion, which succeeded for a few days until troops arrived. Its aftermath saw the introduction of more repression, prohibiting the education of blacks, but also a debate about slavery's future in the Virginia legislature that narrowly failed to introduce gradual emancipation. Revolts on slave ships were widespread, with almost five hundred recorded cases, including the *Amistad*, whose captive-turned-rebel crew was freed by an 1841 Supreme Court case defended by former president John Quincy Adams.

The most successful uprising was the 1791 revolution that established Haiti as the second independent nation in the Americas (after the United States). In South and Central America, runaways formed free communities, preserving their African languages and culture. In Panama, some of these "Cimarron" communities forced Spain to officially recognize their independence. As the largest slaveholding nation, Brazil was home to large free African communities. Slaves would flee into the jungle to build parallel "Maroon" societies, often in close proximity to Portuguese settlements. Some free communities grew into African kingdoms in the Americas. The most enduring was Palmares, which lasted almost a century until its capital was razed by Portuguese forces in 1694. At its height, Palmares supported a free community of over ten thousand, ruled by King Zumbi, who'd descended from Kongolese nobility. Even after Palmares was destroyed, many of its inhabitants managed to escape into the jungle, establishing new settlements.

The reunification of humanity following our primeval diaspora from Africa was a turbulent process, with impacts reverberating to this day. It carried humans around the world, some free, others in chains. It drove revolutions in politics, transportation, economics, and science. It was an unprecedented exchange of people, plants, and ideas that created the modern world. It was in our nature to push our boundaries beyond the known, forging a worldwide network that collected the entire planet into its embrace. Connecting the world caused undeniable tragedy and cruelty, but it also drove technological and social change that improved the lives of billions. The short-term impacts of human reunification were often catastrophic, but in the long run, the indomitable human resolve to understand differences, heal old wounds, forge new relationships, and share ideas can't help but brighten our prospects for the future.

14 | OPENING CONTINENTS

The world had been circled and connected by trade routes, but the dream of reaching Asia by sailing west wouldn't die. Magellan had found a route, but it was too remote to be practical. Might alternate routes exist near the Arctic* or through the continents? No sooner had Columbus returned from the New World than a search for a northern route began. Over a dozen Spanish, Portuguese, French, Danish, Dutch, and British explorers kept trying, and many perished in this quest to make geography conform to human desires. As it gradually became apparent that no such passage existed, the New World settlements became ends in themselves, with new purposes driven by their unique goals and aspirations.

Europeans thought that a passage to Asia could exist as either an ice-free route over Canada or possibly a river route through the continent.† Jacques Cartier's 1535 expedition up the Saint Lawrence that claimed Québec for France was motivated by such a search—and he supposed the "Lachine" rapids at Montreal to be

---

* Europe and Asia are actually remarkably close via the "great circle" polar route. Airplanes flying between North America and Europe usually take this route over Greenland. Britain is closer to Japan than to South Africa.

† This was based on two false assumptions. First, Europeans assumed that salt water couldn't freeze, so there should be an ice-free ocean passage. Second, they assumed that a river could somehow wind its way through a continent. This betrays a dim understanding of hydrology, because rivers must by definition flow downhill, propelled by gravity.

the only thing barring his way to China (in French "China" is "La Chine"). The 1607 Jamestown colonists were unceremoniously dumped in Virginia with instructions to either conquer the natives, as the Spanish had done, or trade with them for gold, but their long-term objective was to follow a river through the continent to the Pacific, and thence on to Asia. Even America's largest city was founded in the aftermath of an attempt to follow a river through the continent. In 1609, the Dutch hired Henry Hudson to sail up the river that now bears his name as far as Albany, giving them claim to New York.

Hudson returned to England, presented his log to the Dutch ambassador, and decided to make another attempt to find a passage in 1610. Sailing this time on behalf of England at the helm of *Discovery*, he skirted south of Greenland into the Canadian Arctic. Passing into what's now known as the Hudson Strait, *Discovery* rounded Labrador in August in great excitement. In front of them was what appeared to be a vast open ocean, leading the way to Asia. What they'd actually stumbled into was a giant inlet twice the size of the Mediterranean—today known as Hudson Bay. Hudson and his crew explored the bay for the rest of 1610 without any luck, and as winter closed in, *Discovery* was frozen into the ice. In the spring of 1611, Hudson vowed to continue probing for a passage, but by June the crew decided they'd had enough and mutinied.

We only know what happened next through the survivors, who were arrested upon returning to England. They admitted to putting Hudson and eight followers, including his teenage son, in *Discovery*'s boat and setting it adrift. The boat broke out oars and rowed to catch up, but was soon left behind as *Discovery* sailed for home. According to testimony, the mutiny was instigated by men who perished in the return voyage—suspiciously convenient, but the survivors were spared thanks to their valuable geographic knowledge. Although Hudson had disastrously failed to reach Asia, he did explore a vast area of the Canadian Arctic. The bay named in Hudson's honor soon became an important route for fur traders traveling deep into the continent. As the spice trade declined, the fur trade boomed, supplying fashionable Europeans with warm hats made from the creatures of North America.*

While most British colonists concentrated in the thirteen colonies, the French forged alliances with Native Americans, traveling vast distances through the continent with canoes and native guides. This pattern was driven by demographics: there were fewer colonists in New France, especially women—despite efforts to recruit orphan girls off the streets of Paris to ship over. Thus, French pioneers extensively intermarried with native women. Some British explorers adopted the French pattern of using canoes to follow rivers, especially in Hudson Bay. The most prominent was Alexander Mackenzie (actually Scottish), who in 1789 discerned from Native Americans the existence of a large river flowing from the continent's interior in a northwesterly direction. Since all other known rivers flowed into either the Gulf of Mexico, the Atlantic, or Hudson Bay, Mackenzie wondered if this might, at long last, flow to the Pacific. He decided to trace the river by canoe to its mouth.

This river, part of the second-longest river system in North America, actually turned out to flow into the Arctic. His hopes of reaching the Pacific dashed, Mackenzie dubbed the waterway "Disappointment River" (although, perhaps to his posthumous dismay, it was later renamed in his honor). Not willing to give up, three years later he set out once again, accompanied by two native guides, six French-Canadian voyageurs, his Scottish cousin, and a dog who seems not to have had a name (referred to in his journal simply as "our dog"). This time the group paddled upriver to the west, eventually meeting some natives who described a trail that wound up into the Rocky Mountains and then down to an ocean beyond. Following this trail, they came to a river flowing straight west out of the mountains. Descending the Bella Coola River, Mackenzie and his party became the first

* In 1670, the Hudson's Bay Company was established to capitalize on the fur trade. North America's oldest corporation, today it runs department stores like Hudson's Bay (Canada), Home Outfitters, Lord & Taylor, and Saks Fifth Avenue.

Europeans to cross North America and reach the Pacific north of Mexico.*

The best-known transcontinental expedition followed a similar pattern to Mackenzie's. In 1803, a young United States concluded the Louisiana Purchase with Napoléon's France, spending only $15 million to double its territory.† President Thomas Jefferson conceived the Lewis and Clark expedition partly as a survey of the newly acquired land, partly as a meet-and-greet with Native Americans who now lived under the American flag, partly as a search for a passage to the Pacific through the continent, partly to stake a claim on the Pacific Northwest, and partly as a scientific expedition to report on the region's climate, plants, and animals. Jefferson was a fossil enthusiast, and he urged Lewis and Clark to take special care to look out for mammoths, which he assumed still roamed the plains. If they'd found any, we might today have a mammoth instead of an eagle as the national animal of the United States.‡

To lead the "Corps of Discovery," Jefferson handpicked his aide Meriwether Lewis, an amateur scientist and former army captain, but Lewis elected to share the responsibility with his former superior officer William Clark. Although joint command can be chaotic, this partnership turned out to be among the best in history, blending Lewis's energy and intellect with Clark's cool and steady demeanor. Lewis prepared for the journey by studying botany, cartography, and navigation with the top scientists in the country. The physician Benjamin Rush supplied remedies such as "Dr.

---

* Surprisingly, they weren't the first Europeans to reach the outlet of the Bella Coola River—in fact, they'd just missed the first seaborne expedition to the area by forty-eight days. This group had been commanded by George Vancouver, who sailed along the Pacific coast of Oregon, Washington, and British Columbia (from across the Indian Ocean and Pacific after sailing around Africa).

† Around $250 million today, or forty-five cents per acre. Typical farm real estate is six thousand times higher, making this one of the best deals in history.

‡ Jefferson's fascination with mammoths was partly motivated by European claims that American plants, animals, and people were somehow smaller and inferior. This may be the origin of the American obsession with having the biggest version of everything—exemplified by the "Mammoth Cheese" incident, in which the town of Cheshire, Massachusetts, presented Jefferson with a 1,234-pound block of cheese made from the milk of every cow in town.

Rush's Thunderbolts," concoctions of chlorine, mercury, and a root blend—purported cure-alls, but likely to just have an extreme laxative effect. (They might not have cured you, but they'd get you out of bed.) Preparations complete, Lewis traveled to St. Louis (recently acquired from France), where he met Clark and thirty men.

The expedition set out in early 1804, following the Missouri River into the continent's interior. Instead of rowing against the current, the men hauled their supply-laden shallow-bottomed craft upriver using ropes from the riverbanks. Along the way, they met Native Americans and distributed specially prepared silver medallions of friendship that were simultaneously intended to convey the message "you are now part of the United States." Although the expedition was peaceful, it was also well armed. Only once did violence break out, when Blackfoot warriors tried to steal some guns from the camp under the cover of night (two were killed in the resulting skirmish). The expedition also narrowly avoided conflict with Spain (which claimed some territory along their route). A Spanish force of fifty soldiers dispatched from Santa Fe tried to arrest the Corps of Discovery, only to learn that they'd already passed the intercept point in Nebraska a few weeks prior.

Following the Missouri River through what is now Kansas City and Omaha, the expedition passed into the Great Plains, a plateau teeming with deer, elk, and bison.* Here they encountered prairie dogs, which dig extensive tunnel systems and use a language of barks and chirps to warn of danger. Jefferson had requested samples of flora and fauna, so Lewis sent back a live prairie dog; it went on to live for many years in a Philadelphia exhibit. As winter approached, the Corps of Discovery erected a camp called Fort Mandan in North Dakota, around the halfway point. This attracted Native American visitors curious to learn the identity of the newcomers. To further warm relations, Lewis and Clark agreed to visit the local Mandan camp, where they passed around a ceremonial pipe. It was here that they encountered a remarkable young woman.

* Bison once roamed the plains in tens of millions, but hunting reduced their population to just 1,091 by 1889. Under protection, they've rebounded to around 500,000 today.

Only seventeen at the time, Sacagawea had been captured five years earlier and sold as a wife to the French fur trapper Toussaint Charbonneau, who was visiting the Mandan village. Since Sacagawea had come from the west, Lewis and Clark decided to hire Charbonneau—not for his skills but for those of his young wife, who could act as interpreter. Sacagawea soon proved her worth, most crucially in using negotiation savvy to defuse tensions with Native Americans along the way. Even her very presence was useful, since a young woman carrying an infant strongly signaled that the expedition was a peaceful venture instead of a war party.*

As the snows cleared in April 1805, the Corps of Discovery headed west through Montana. Sacagawea started recognizing landmarks from her childhood as the group traced the Missouri River to its headwaters and hiked up trails through the Rockies to the Continental Divide. On several occasions they encountered grizzlies, far larger than bears back east. Seaman, a Newfoundland dog, kept lookout and warned of danger: at the approach of bears, he'd tense up and bark. By August, the expedition was running low on food and becoming disoriented in the mountain passes. Lewis was scouting ahead with three men when he came across some natives of the Shoshone tribe who'd never met Europeans or Americans. Hoping to foster friendly relations, he presented them with gifts and made peaceful signs he'd learned. The Shoshone seemed to understand but were still apprehensive.

For four days, Lewis waited. The Shoshone had horses, which he wanted to acquire for the trek through the Bitterroot Mountains to the Pacific. When the rest of the group arrived, Sacagawea came forward to translate, and suddenly, in an amazing coincidence, recognized that she'd made it home. The Shoshone chief was her brother. The tribe provided the expedition with horses and instructions on how best to reach the Pacific. Such was Sacagawea's dedication that, instead of remaining with her family, she decided to accompany Lewis and Clark the rest of the way. Passing through

* The infant, Jean Baptiste Charbonneau, was adopted and raised by William Clark after Sacagawea died six years later. He became a prominent American explorer and military scout, speaking more than half a dozen languages, including English, French, Spanish, German, and several native tongues.

the mountains, they paddled canoes down a series of rivers, eventually leading to the Columbia. As they followed the river, they encountered natives wearing British sailor caps—a sign the Pacific was at hand.* Soon they caught glimpses of the great ocean.

They'd reached the Pacific at last, but winter was setting in. The expedition held a vote on where to erect a winter camp. Each member voted, even Sacagawea and a slave named York—perhaps the first instance of a woman and black man voting in America. And thus it was that a fort was built near Astoria, Oregon, the first time an American flag had been flown on the Pacific coast.

Food ran short as the winter turned bitterly cold, and the expedition was by now too poor to trade with local natives. As the Corps of Discovery waited, they scanned the horizon for passing ships, which by this time regularly visited the area to trade for otter furs. Jefferson had given Lewis a note promising payment to any ship that would return the expedition by sea, but by March 1806 no ship had been spotted. Setting out overland, the group made excellent time on the return journey. By late September they spotted the first sign of civilization: a cow grazing peacefully by a riverbank. A few days later, they entered St. Louis and reported their story to an enthusiastic nation.

In the strictest sense, we can't really say that Lewis and Clark discovered anything. All the regions they entered were known to Native Americans, and three-quarters had already been visited by Europeans. Rather, the expedition's contribution was consolidation and dissemination. Previously, geographic knowledge had been compartmentalized and largely unrecorded. The first Europeans to visit the west were fur trappers, and it didn't make sense for them to disclose sensitive commercial information. The Lewis and Clark expedition compiled detailed maps and cataloged the climate, terrain, plants, and animals of the region, sharing their observations with the world.

---

* British, French, Russian, Spanish, and American ships had been visiting the Pacific Northwest for two decades. The region was jointly claimed for Britain by George Vancouver and for the United States by Robert Gray, who circumnavigated the world in 1787–1790. The joint ownership dispute was resolved in 1846, whereby the United States kept Washington, Oregon, Idaho, Montana, and Wyoming, and Britain retained British Columbia—i.e., the British portion of the Columbia River.

As impressive as Lewis and Clark's accomplishments were, in many ways the two explorers' exertions were surpassed by those of lesser-known men who trekked across Siberia. We may wonder how Russia came to be the largest country on Earth. In fact, the story is similar to the American one, the main difference being that Russia expanded into twice the area in half the time, two centuries earlier. This was, without a doubt, the most impressive permanent national expansion ever. Between 1450 and 1850, Russia grew an average of thirty-five square miles per *day*—equivalent to adding an area the size of Belgium every year, for four hundred years. After throwing off Mongol rule, Moscow spent the 1400s unifying the Russian principalities, and by 1556 it had defeated the Mongol successors of Kazan and Astrakhan, opening the way to Asia. From there, explorers reached the Pacific in less than a century (1639) and built the first settlement on the Pacific in 1647. The next year, the Cossack Semyon Dezhnyov sailed from the Arctic into the Pacific through the Bering Strait between Siberia and Alaska. In 1689, Russia became the first European nation to establish diplomatic relations with China.

The most impressive Russian expeditions were led by the Danish explorer Vitus Bering. In 1725, Bering set out from St. Petersburg with thirty-four men across the vast expanse of Siberia. Covering twice as much distance as Lewis and Clark in a grueling two-year overland trek was merely the beginning. Reaching the Pacific, they built a ship and then sailed to chart the coastline of Alaska. In American terms, this would be equivalent to Jefferson ordering Lewis and Clark to march twice across the continent and then build a ship to explore Japan. During two voyages, Bering claimed Alaska for Russia and established that Asia and North America aren't connected but separated by a mere fifty miles of sea.* Subsequent explorers would build Russian forts as far south as California—closer to Australia than to Moscow or St. Petersburg.

---

* Russia sold Alaska to the United States in 1867 for $7.2 million. Ridiculed at the time, this deal seems in retrospect to be among the best in history, along with the Louisiana Purchase (British envoys declined to purchase Alaska, which could have gone to Canada instead). Imagine how different the Cold War might have been had Russian bases been located within five hundred miles of Seattle!

Nor was the Corps of Discovery the only exploratory venture commissioned by Thomas Jefferson. In 1806, the Red River Expedition traced the eponymous river from Louisiana over six hundred miles into Texas; there the group was turned back by Spanish troops. In the same year, Zebulon Pike led a party across Kansas and Colorado. Due to a geographic error, he took the wrong pass through the Rockies, getting captured by the Spanish. This turned out to be a bonus, because as Pike was led through Spanish New Mexico, Mexico, and Texas, he compiled detailed notes on what he saw. At the time, Pike's travels were almost as famous as Lewis and Clark's, and his journal was translated into half a dozen languages. Combined, these expeditions culminated a three-century effort as explorers from the east met up with others coming across Mexico and the Pacific. Half a century before the transcontinental railroad, the great blanks in the map had finally been filled.

15 | FRONTIERS OF SCIENCE

Patterns of exploration are influenced by changes in society, and as the Age of Enlightenment dawned in the eighteenth century, it swept away old orthodoxies in favor of new ideas about liberty, reason, and tolerance. These were the ideas that sparked the American and French Revolutions, the political branches of the scientific revolution. Driven by enlightenment ideals, exploration gradually evolved from the wholesale plunder of foreign lands in search of gold to more virtuous scientific purposes. Instead of conquistadors, explorers were now botanists, physicists, astronomers, and anthropologists. It was no longer enough to merely find the world—exploration now meant truly discovering it.

Curiosity had always been an exploratory motive, but an international competition for prestige based on scientific discovery was something new. Even the ships were intended to reflect the new ideals under which they sailed, with names such as *Discovery, Resolution, Endeavour, Adventure, Géographie, Naturaliste*, and *Astrolabe*. Some ventures, such as the French 1735 mission to Ecuador to measure the shape of Earth, were virtually devoid of nonscientific purpose.* Nevertheless, even purely scientific voy-

---

* The idea was to measure a degree of latitude near the equator and compare it to a degree up in the north. This would determine if Earth bulges at the equator and flattens at the poles (it does).

ages were intended to enhance the image of the sponsoring nation in an eighteenth-and-nineteenth-century equivalent of the Cold War space race. In this exploratory contest, a new figure emerged: the scientist-hero, conquering ignorance on behalf of the nation.

These scientific voyages were themselves enabled by new technologies. Latitude had long been measured from the angle of the Sun, but no means of determining longitude existed until the mid-1700s, when the clockmaker John Harrison invented an accurate chronometer. If you synchronized the chronometer to the time in the location from which you departed, a comparison of its time with the Sun-based local time told you how far you were around the planet.* Most ships carried multiple chronometers as backups (Charles Darwin's HMS *Beagle* had twenty-two) and synchronized them in port with balls dropped at specified times (origin of the New Year ball drop in Times Square). Longer voyages were enabled by improved nutrition. While Spanish and Dutch captains had previously charted courses through the Pacific, they could never stay long before scurvy turned their ships into floating hospitals.

Astronomers knew that by observing the transit of Venus from multiple points on Earth, they could compute the distance to the Sun. Edmund Halley had predicted that Venus would pass in front of the Sun in both 1762 and 1769 but then not again for another century. Elaborate preparations were made to observe the 1762 transit, but overcast weather spoiled most of the data. The scientific community decided to take no chances for 1769. This effort became one of the most anticipated scientific collaborations in history, coordinating measurements across the globe. To lead the Pacific contingent, the British Admiralty would need a captain with a broad range of skills: navigator, cartographer, astronomer, and commander. They found all of these in James Cook.†

---

* Imagine you set your watch in New York and fly to Los Angeles. If your watch reads three p.m. but you measure the Sun at high noon, you know you're three twenty-fourths of the way around the world.

† Captain James Cook was an influence for Captain James Kirk of the starship *Enterprise*, with his "to boldly go where no man has gone before," an allusion to Cook's "to go not only farther than any other man, but as far as it was possible to go."

To prepare for the expedition, Cook had the navy retrofit an old coal-hauling barque. Not fast or much to look at, *Endeavour* was practical: sturdy with a large cargo capacity and able to sail into shallow waters or beach herself to make repairs.* *Endeavour*'s purpose was clear as it sailed for the South Pacific with fourteen scientists led by Joseph Banks, the premier botanist of his age. For such a long voyage, nutrition was key, so Cook took a page from the Vikings and packed *Endeavour*'s hold with vats of sauerkraut, dried soup, and carrot preserve. This was the first expedition with a varied diet, with remarkable results. While it had been customary for ships to lose dozens of men to scurvy during long voyages, *Endeavour* had not a single case. The voyage to Tahiti went smoothly, and the astronomical observations of the Venus transit were a global success.

Endeavour was actually the fourth European ship to visit Tahiti. Two years prior, competing French and British expeditions had raced across the Pacific, stopping at Tahiti within a few months of each other. The British ship *Dolphin* arrived first (carrying a goat mascot that later accompanied Cook, becoming the first nonhuman creature to circumnavigate the world twice). The two-ship French expedition was commanded by the eminent scientist-adventurer Louis-Antoine de Bougainville, who lent his names to islands, plants (bougainvillea), and thirteen vessels of the French navy. A prodigy, Bougainville was inducted into the British Royal Society for his treatise on integral calculus, which he published at age twenty-five. In 1759, he and Cook came within a few miles of each other during the Battle of Québec.†

The British captain Samuel Wallis described the Tahitians as exceptionally friendly and must have raised eyebrows in the stodgy Admiralty with his descriptions of beautiful island women.

---

* Space Shuttle *Endeavour* was explicitly named after Cook's ship, hence the British spelling.

† As aide-de-camp for Montcalm in 1757, Bougainville negotiated the surrender of Fort William Henry near Albany, for which he has a cameo in the 1992 movie *The Last of the Mohicans* (details in the movie are largely based on Bougainville's fabulous diary). In 1781, he was instrumental in defeating the British fleet around Yorktown during the pivotal battle of the American Revolution, and in 1799 was appointed senator by Napoléon.

Bougainville was more direct, naming the island "New Cythera" after the birthplace of Venus (goddess of love). This voyage caused a sensation in Europe when it emerged that the ship's surgeon had snuck his mistress on board (the first woman to circumnavigate the globe). The Tahitians were so enthralled at meeting a European woman that they lavished gifts on her and composed eulogies in her honor. The philosopher Jean-Jacques Rousseau used this encounter to promulgate his "noble savage" concept, whereby "primitive" people who are stripped of the corrupting influences of civilization reveal the basic goodness of humanity.

To prevent incidents involving *Endeavour*'s crew, Cook ordered that only official representatives of the ship be allowed to go ashore. Relations were amicable, although Cook worried that the Tahitians were being corrupted by European ideas. Later, Cook would make similar observations of Aboriginal Australians, remarking that they were "far happier than we Europeans; being wholly unacquainted not only with the superfluous but the necessary conveniences so much sought after in Europe, they are happy not knowing the use of them. They live in a tranquility which is not disturbed by the inequality of condition." Cook's insight captured a key aspect of the human psyche. Material wealth does not make us happy. Instead of finding happiness in modern conveniences, we compare our station to our neighbors and, finding ourselves not at the top of the heap, remain unsatisfied.

After the transit of Venus and a well-deserved rest in Tahiti, Cook set out to follow his secret orders. His mission was to continue south, searching for a predicted "Terra Australis Incognita." This great continent was thought to exist in the South Pacific, bringing balance to the globe, without which Earth would tilt in space. Australia was already known but not thought large enough to balance the world, and no European had been far into the South Pacific. The blank spot on the map was large enough to hide another continent at least the size of Asia. *Endeavour* set out due south, but after almost two thousand miles of sailing through open ocean, no land was sighted. Following instructions, Cook turned toward New Zealand (discovered by Abel Tasman in 1642) to discern if it was perhaps a peninsula of the great continent. It wasn't, so Cook charted its coastline to establish that New Zealand is composed of two islands.*

Since *Endeavour* was on the way home, Cook decided to chart the outline of Australia (then "New Holland") as he sailed. In April 1770, the ship dropped anchor in Botany Bay (near Sydney), named by the ship's lead scientist, Joseph Banks, for its wondrous flora and fauna. Cook reported that Australia seemed like nice rolling country, sparsely inhabited based on encounters with Aborigines and smoke signals rising from the shore. He took a special liking to kangaroos, describing them as both fascinating and delicious. While sailing along the northeast coast, Cook remarked on the incredible expanses of coral (part of the 1,500-mile Great Barrier Reef), which nearly spelled disaster when *Endeavour* ran aground on a shallow patch. Most ships would have been torn apart by the swelling seas and sharp coral, but *Endeavour* was sturdy, and lucky, too. Cook ordered everything not nailed down to be cast overboard, and after the crew feverishly pumped and used chains to drag the hull back into deeper water, *Endeavour* broke free.†

After stopping on a beach to patch the hull (at a place now called Cooktown), *Endeavour* limped to Dutch Indonesia to make repairs. A few months later, Cook and his men headed home, arriving in England three years after they had left. Cook's first voyage had been a resounding success, but there still remained a big blank on the map where a great southern continent could lurk. A second voyage was planned, focusing entirely on a search through the world's southern oceans. It would be dangerous and monotonous, the longest ever out of sight of land. Joseph Banks had intended to go but changed his mind before sailing, so he was replaced by Johann Forster and his son Georg (who would become a famous naturalist in his own right). Two ships were outfitted in the same manner as *Endeavour*, and in 1772, Cook set out for the South Pacific again, this time commanding *Resolution* and *Adventure*.

On his second voyage, Cook circumnavigated the globe in the

---

* The strait separating the islands is now named for Cook, and *Endeavour* is featured on the fifty-cent New Zealand coin.

† *Endeavour*'s discarded cannons were recovered in 1969 and are now displayed at a Sydney museum.

opposite direction, farther south than any ship had previously sailed, through seas thick with ice and fog. Along the way, he returned twice to New Zealand and once to Tahiti. As he filled in blanks on the map, Cook discovered many southern islands but forever disproved the idea of a great southern continent needed to keep Earth in balance. By default, Australia inherited the title of "Terra Australis," but it was no longer incognita. Although he never sighted Antarctica, Cook predicted its existence based on rocks he noticed buried in icebergs, indicating they'd come from a southern land of ice and snow.

Cook's second expedition returned in 1775 after three more years at sea, but an experienced captain such as Cook was too valuable to leave on shore for long. The Admiralty had a new mission, one that had been the holy grail of exploration for centuries: find the Northwest Passage. Dozens of explorers had tried and failed, but this time Cook would force his way through the Arctic from the Pacific at the Bering Strait near Alaska. This would also fill in the last big blank patch on the map, along the Pacific Northwest. The American Revolution had just broken out, but Benjamin Franklin convinced the fledgling American navy that Cook's mission benefited everyone. If encountered, he should be treated as a friendly ship and given aid. The French, Spanish, and Dutch, who joined America in the war against Britain, followed suit in pledging Cook assistance. For the first time in history, it was universally acknowledged that an explorer was sailing on behalf of the entire world.

Cook departed England on July 12, 1776, eight days after the Declaration of Independence. He again commanded *Resolution*, this time accompanied by a new ship, *Discovery*.* Rounding the Cape of Good Hope, the ships stopped briefly in New Zealand and then sailed to Tahiti. From there, they headed due north toward the Bering Strait, but along the way stumbled into Hawaii. Although Spanish galleons sailing between Manila and Acapulco had passed nearby for centuries, no European had yet encoun-

* Along with Cook was William Bligh, of later *Mutiny on the Bounty* fame.

tered the islands.* After a brief stop, Cook sailed for the American coast, landing in present-day Oregon, just south of the Columbia River. He then proceeded up the coast toward Alaska, all the while searching for a passage that might lead to the Arctic, questioning Native Americans along the way. Charting the entire northwest coast from Oregon to Alaska, Cook eventually reached the Bering Strait, which was already known from Russian reports.

Feeling his way through the Bering Strait, Cook was confronted with an impenetrable wall of ice. Since winter was approaching, he sailed back to Hawaii, intending to resume the search in spring. Cook described the Hawaiians as the most welcoming people on Earth, and they plied his crew with tropical delights. However, tensions mounted as the crew continued to eat the island's food and seek the company of its women. Rumors spread that the ravenous Europeans had been driven from their lands by famine and were now consuming all the food in Hawaii. Brawls erupted, and British marines fired into angry crowds. Cook went ashore to pacify the situation, but as he turned to his boats on the beach, he was clubbed to death by the Hawaiian mob.† Dispirited, the *Resolution* and *Discovery* crews sailed away without their captain. After one last search for an ice-free passage through the Bering Strait, they turned for home.

With Cook, the outlines of continents had been established. As the eighteenth century gave way to the nineteenth, a new breed of explorer tackled the great scientific questions of the day, probing the nature of Earth, its climate and features, and its people, plants, and animals. The diversity of life observed by Darwin on HMS *Beagle*'s 1831–1836 circumnavigation of Earth inspired his theory of evolution. But there was an earlier scientist whom Darwin looked to as his biggest influence—a genius who advanced fields

---

* Cook named Hawaii "the Sandwich Islands" after the lord of the Admiralty (the fourth Earl of Sandwich), who invented putting meat between slices of bread so he could eat while playing cards, hence "sandwiches." Hawaii's British connection is preserved by the fact that it remains the only state flag featuring a Union Jack.

† There is confusion over how the Hawaiians perceived Cook—if not as a god, then at least as a powerful chief. His bones were preserved as relics, with some returned to his crew for burial at sea.

across the spectrum: biology, climatology, ecology, geography, and geology. In fact, his broad range of contributions is probably why he's not better known today. Driven by a desire to understand the interconnections underpinning everything, Alexander von Humboldt was the world's preeminent scientific traveler.*

Across the square in Berlin where Nazis burned piles of books, a statue sits before Humboldt University dedicated to the "second discoverer of Cuba." Accurate though this description may be, Alexander von Humboldt was much more. By the late 1700s, Europeans had known the Americas for centuries, but in many ways they were still lands of mystery. Spanish authorities had closed their colonies to outsiders, making no effort to survey them from a scientific perspective. Humboldt had grown up fascinated by tales of exploration. A minor Prussian noble, he studied geology and took a government job that allowed him to travel around Germany, as director of mines. To whet his appetite for adventure, he had Georg Forster introduce him to Joseph Banks, who captivated Humboldt with stories and vast collections of plants and animals from Cook's voyages. Humboldt decided then and there to reject the life of a German bureaucrat to travel the world.

He'd inherited some money, so Humboldt could finance his own trip. He wrote the Spanish king for permission to visit the American colonies, which, to his surprise, was granted. His journey through the Americas (1799–1804) was the region's first scientific survey. Disembarking with a mountain of scientific instruments, Humboldt and his French companion, Aimé Bonpland, proceeded to lug thermometers, barometers, rain gauges, and myriad other contraptions through the jungles of South America. Ever curious, they sampled poison used by natives to tip blow-darts (only lethal when taken intravenously) and drank the "milk" of a rubber tree (vomiting rubber balls for hours). The adventurers were viscerally excited by discovery, to the extent that Humboldt worried Bonpland "might go mad if the wonders do not cease." Their

* Humboldt was considered the second-most famous person of his time (after Napoléon), with thirty-nine US towns and geographic features named after him. Nevada nearly was, and Edgar Allan Poe dedicated his last major work, *Eureka*, to him.

19,286-foot ascent of Chimborazo in Ecuador was the highest anyone had ever climbed a mountain, a record unchallenged for thirty years. Their collected samples doubled the number of known plant and animal species in the Americas, but these represented only a tenth of those they observed. Interviewing native people, they systematically cataloged local societies for the first time.

Humboldt would continue traveling, as in 1829, when he journeyed through Russia and Central Asia to the border of China, but most of the rest of his time was devoted to sharing his observations. He wasn't afraid to aim high, with his five-volume masterpiece *Kosmos* envisioned as a "total natural history of the universe." *Kosmos* became the first widely read scientific book, laying the groundwork for Carl Sagan's 1980 book and television series of the same name.* The first modern science communicator, Humboldt brought the world home to everyday people. Ralph Waldo Emerson called him a "wonder of the world," while to Simón Bolívar, he was "the true discoverer of America." Humboldt considered himself a "universal citizen of humanity."

The nineteenth century also saw the dawn of tourism. An Austrian friend of Humboldt's, Ida Pfeiffer, was perhaps the first woman to travel the world on her own, and certainly the first to make a living by writing about her adventures. An early pioneer of women's rights, she was raised in a large family, where she insisted on being educated just like her brothers. Restless, but forced to marry an older man, she raised two sons and supported her family giving music lessons—but she always yearned to do more. After her husband died and sons moved away, she set out in 1842 at the age of forty-five to explore the planet. Designing her own explorer's outfit, with a skirt, shawl, mosquito net, and broad-brimmed hat, she trekked through the jungles of the world armed with knife and pistol. The dangers were real: one time she fought off a bandit at knifepoint, while another she joked with some cannibals that she was too old and tough to eat.

With each voyage, Pfeiffer wrote a book and used the profits to fund her next trip. Unlike previous explorers, she focused on telling the stories of the people she encountered, including details

* Later adapted into a 2014 version starring Neil deGrasse Tyson.

of home life and women. Her travel works became international bestsellers, translated into many languages, and soon rulers were inviting her to tour their countries to encourage others to visit. It's not an exaggeration to say that Ida Pfeiffer inaugurated the global tourism industry. Whereas Humboldt had introduced people to the world's scientific wonders, Pfeiffer brought home the idea of foreign travel. Now, for the first time in history, ordinary people could explore the world for themselves.

16 | LANDS OF ICE AND SNOW

Exploration begets more exploration, and the unknown calls as a siren to our souls. Even disasters motivate us to follow paths blazed by the first pioneers, attempting to succeed where others have failed. Nowhere was this truer than the Arctic, where repeated catastrophes forged legends of heroic adventure. Mercantile objectives aside, the conquest of the north became a battle of humans against nature. During the Victorian age, a confident European society decided to harness newfound industrial technologies to prevail at last. With Britain victorious against Napoléon, the Royal Navy's energies could be redirected to peacetime struggles. With careful preparation, and armed with modernity's latest marvels, how could such efforts fail?

Sir John Franklin was selected to lead the 1845 expedition to finally breach the Northwest Passage. Franklin was no stranger to the Arctic—he'd led three expeditions there already. His 1819 voyage ended badly when half his crew starved, earning him the moniker "the man who ate his boots." When asked whether his advanced age of sixty was an obstacle, he famously retorted, "I beg your pardon? I'm only fifty-nine." Two warships were specially retrofitted for the 1845 effort: *Erebus* and *Terror*, thick-hulled mortar ships, with iron plating, steam-powered heaters, desalination machines, and locomotive engines driving seven-foot propellers. Both ships had previously distinguished themselves in Arctic and Antarctic waters and had been immortalized in the US national anthem as the vessels firing "bombs bursting in air" (mortars and ship-fired rockets) at Fort McHenry in 1814. The expedition was the first to carry a camera, and the ships were packed with a three-year supply of canned food.*

Erebus and *Terror* sailed† from England in May 1845 with 129 souls aboard. In July, they passed a whaling ship off Greenland and then were never seen again. What happened? Over the years, we've been able to reconstruct their fate, with wrecks of the vessels found in the Canadian Arctic in 2014 and 2016. The crew made good progress in the first year and overwintered at Beechey Island. Pressing on through 1846, the ships became locked in ice during the next winter, and a thaw never came. According to a note later recovered, Franklin died in June 1847, and then a third winter set in. By the spring of 1848, with the ships trapped and food dwindling, the surviving crew decided to trek overland to safety. Based on corpses later found strewn across the Arctic tundra, they didn't make it far.

After two years with no word, the first major international search was organized, with a reward equivalent to $2 million for information leading to the crew's recovery. Dozens of vessels participated, even from rivals such as France and the United States, and at least five ships were themselves lost during the search.‡ Metal buttons stamped with locations of supplies were distributed to the Inuit, hoping they'd reach survivors. In 1854, an Inuit told a surveyor that forty white men had starved in the region a few years earlier, showing him cutlery, coins, books, and other relics from the doomed ships. The rescue was called off, but Lady Franklin commissioned a follow-up search that recovered two messages. The first, dated May 1847, described the expedition's progress and confidently reported, "Sir John

---

* Canning was invented during the Napoleonic Wars to supply troops on distant battlefields, but can openers were not invented until 1855. For the first fifty years, cans were opened with a hammer and chisel.

† The ships had both sails and engine-driven propellers.

‡ In appreciation of American assistance, Queen Victoria presented a desk to Rutherford B. Hayes made from the timber of HMS *Resolute*, one of the search vessels. It now sits in the Oval Office—the president's desk.

Franklin commanding, All Well." The second, dated almost a year later, had a very different tone. Scratched around the margin of the same paper in shaky writing, it reported that *Erebus* and *Terror* had been trapped in ice for two winters, and many had perished. The next day, the survivors would set out overland to try to reach safety.

Several things were strange about the expedition's fate. The last message contained several errors, including spelling and dates. Additionally, abandoned sleds were found loaded with heavy books, silverware, and other items of dubious value in a crisis, suggesting that the crew may not have been thinking clearly. Records before the expedition's departure show that the food cans were sloppily assembled in haste with lead solder "dripped like candle wax." Thus, cognitive impairment due to lead poisoning is a suspect. Also, since pasteurization hadn't been invented, many of the cans would have been contaminated, making the food unsafe to eat. The entire episode was a reminder of the power of nature and reflected the hubris of the Victorian age. While modern technology dramatically failed to conquer the north, the expedition was surrounded by Inuit thriving in precisely the same environment.

Despite the danger, or perhaps because of it, people continued to dream of conquering the north. The Arctic enterprise evolved into a race against rivals for fame and glory. The Norwegian explorer Roald Amundsen was first to wind his way through the ice from Greenland to Alaska, in a three-year voyage starting in 1903. His secret to success was to do the opposite of Franklin, traveling light and living off the land. Amundsen's crew learned valuable survival skills from local Inuit, hunting seals and other animals to acquire fresh meat with vitamins to stave off malnutrition. Dogsleds were used for overland transport, and bulky parkas were discarded in favor of warm and waterproof animal skins.

The pole itself was conquered surprisingly recently. In 1888, a team of six Norwegians and Finns crossed Greenland on skis. The Finns were hired as reindeer herders, with a plan to haul supplies on reindeer-pulled sleds. Unfortunately, unlike domesticated reindeer back home, the wild variety stubbornly refused to pull sleds. The team had to pull the sleds themselves, but since the men were expert skiers, they nevertheless succeeded in crossing Greenland. The leader of this expedition, Fridtjof Nansen, next set out to reach the pole. In a specially designed vessel, *Fram*, he intended to intentionally get stuck in the ice and simply drift north to the pole. He departed in June 1893, and the first step worked brilliantly. For months, *Fram* drifted ensconced in ice, but Nansen became impatient as the drift rate slowed. By March 1895, he could wait no longer. He set out with a companion to dash by ski across the remaining four hundred miles of ice.

They seemed to be making good progress, but soon Nansen realized that the ice pack was drifting south, canceling their efforts. As food ran low, they were still two hundred fifty miles from the pole, so with a "chaos of ice blocks stretching as far as the horizon," they turned back. As the weather turned colder, Nansen decided to camp for the winter in Franz Josef Land.* Constructing a small shelter, they stocked up on meat from bears, walruses, and seals. In May 1896, they resumed their journey south, and one morning Nansen woke up to the sound of barking dogs. By sheer coincidence, he'd managed to run into the British explorer Frederick Jackson, who was leading an expedition north. Nansen and his companion were rescued and returned to Norway. A few months later, *Fram* emerged from the ice, and the crew reunited in Oslo to national celebration.†

The North Pole remained unconquered, so others lined up to try. In 1909, Robert Peary led a party of twenty-three from Ellesmere Island in Canada toward the pole, hauling sleds along the way. Once they'd reached a point one hundred fifty miles from their goal, Peary and five companions‡ made the final dash on skis. On April 6, 1909, Peary measured that they were within five miles of the pole, but since no one else was trained with the sextant, this

---

* An island off Siberia named after the Habsburg emperor by the 1872 Austro-Hungarian Arctic Expedition.

† In 1922, Nansen won the Nobel Peace Prize for his work reuniting displaced refugees of war, issuing "Nansen passports" that were recognized around the world.

‡ His companions included Matthew Henson, the first African-American Arctic explorer, and three Inuit hunters.

was never confirmed. Additionally, when he returned, errors were noted in his calculations that suggested the team had gotten no closer than thirty miles. To make matters worse for Peary, another explorer, Frederick Cook, also claimed to have returned from the pole around the same time. Peary's supporters quickly organized a smear campaign to tarnish Cook's reputation and discredit his data. Legal battles ensued, and in 1911 the National Geographic Society and Naval Affairs Subcommittee of the US House of Representatives ruled in favor of Peary—but this ruling may have been biased, since Peary's expedition was supported by both organizations. Peary won the popularity contest, but it's unclear if either man actually reached the pole.

If not Peary or Cook, then who was first? In 1926, Admiral Richard Byrd piloted a plane in a fifteen-hour marathon flight over the pole from Spitsbergen, but doubts arose based on the flight duration (too short to cover the distance) and inconsistent testimony from his copilot. Then, just three days later, the sixteen-man crew of the airship *Norge* became the first to definitively reach the pole. The *Norge* expedition was conceived by the Norwegian Roald Amundsen, financed by the American Lincoln Ellsworth, and piloted by the Italian aviator Umberto Nobile. Thus, *Norge* dropped Norwegian, American, and Italian flags as it flew over. Controversy erupted when it turned out that the Italian flags were larger and Nobile and the Italian dictator Mussolini were taking the credit, touting the expedition as proof of fascist Italy's emerging prowess.

Two years later, Nobile hoped to follow up with an identical airship, *Italia*, but this effort became a fiasco when a storm struck and crashed the ship after it reached the pole. Six men from *Italia* were carried off in severe winds, never to be heard from again, while ten survived the crash, stranded on the ice. An international search began, and a month later, the survivors were finally located. A Swedish ski plane swept down to make contact, but it could only carry one man to safety. Nobile wanted to stay with his crew but was talked into being rescued. This turned out badly, since it played out in the press as cowardice. Nobile's reputation never recovered. In an interview, he detailed grievances against Musso-

lini, but the fascists pinned the blame on Nobile, who resigned
from the air force and fled Italy. Seven more crew were eventually
rescued by Soviet icebreakers, but the entire episode was a public
relations catastrophe, with morbid rumors of the Italians' canni-
balizing the Swedish meteorologist. One of the biggest blows was
the loss of Amundsen, whose plane disappeared during the search
for *Italia*'s crew.

The north was conquered for good when a party of Soviet scientists landed near the pole and walked to it in 1948. Ten years
later, the first nuclear submarine, USS *Nautilus*, sailed under the
pack ice from end to end. The next year, another nuclear sub, USS
Skate, surfaced at the top of the world. Nowadays, with modern
technology and warming oceans, the Arctic no longer seems quite
so formidable. Almost every year witnesses a record retreat of sea
ice, and in 2007, the Northwest Passage opened for the first time.
The dream of sailing from Europe to Asia across the top of the
world is now being realized, for at least a few months of the year.
In 2013, China started sailing container ships across the Arctic,
cutting the distance to Europe in half. Not only has the north been
tamed, it's become a tourist destination. In 2016, Crystal Cruises
started offering twenty-eight-day cruises from Vancouver to New
York through the Northwest Passage. Clearly, at least some consider the breathtaking views to be worth the $50,000 evacuation
insurance.

The quest to conquer the Antarctic unfolded in parallel with
the Arctic quest, often with the same explorers. However, unlike
the north, this last landmass on Earth had no apparent commercial value. This was purely a battle of the human spirit. James
Cook had theorized the existence of Antarctica from rocks
trapped in icebergs, but his prediction that no one would ever
set foot there proved a rare misjudgment. Humans have never
been able to leave a frontier unconquered for long. A series of
voyages had landed on Antarctica by the mid-1800s, including
one that involved Franklin's *Erebus* and *Terror* (which lent their
names to Antarctic mountains). Yet, the continent's vast interior,
almost twice the size of the contiguous United States, remained
firmly beyond reach.

Just as the conquest of the North Pole reached its climax, the exploration of Antarctica was becoming an international race. Russians, Germans, Swedes, Norwegians, French, British, Americans, Australians, and others scrambled to reach the remotest point on Earth. In 1898, a Belgian expedition overwintered on the continent, and by the early 1900s, the British explorer Robert Falcon Scott had penetrated deep inland using dogsleds. In 1909 (the same year Peary "reached" the North Pole), Ernest Shackleton led an expedition toward the South Pole using motorized sleds. Bad weather forced them to turn back, but not before they reached the polar plateau within a hundred miles of their goal. The final push was at hand, and in 1911 two rivals set out: a Norwegian party led by Roald Amundsen, and a British team led by Robert Falcon Scott. The race was on.

Scott's team was lavishly equipped, with motorized sleds, ponies, and a prefabricated camp. However, the sleds soon broke down, so equipment had to be hauled across the frozen landscape by human power. The team established their camp 175 miles from the pole, at which point Scott led four men for the final push. Hauling the supplies was exhausting, but after two weeks of trudging, Scott's team finally reached the pole—only to discover that Amundsen had beaten them by thirty-four days! The Norwegian team had traveled light, with sleds pulled by dogs, many of whom they ended up eating. At the pole, Amundsen had left a Norwegian flag, his tent, some supplies, and a letter to Norway's king with a polite request for Scott to deliver it. Scott's demoralized British team slunk back to the safety of their camp, but bad weather soon made the situation much worse.

As Scott's party retreated through plummeting temperatures, they grew weaker by the day. One of Scott's companions, named Oates, injured his hands and developed severe frostbite on his feet. To avoid slowing down the team, he stoically wandered out one night with the words, "I am going outside and may be some time." Oates's sacrifice allowed the rest to travel faster, but soon a massive blizzard hit, just eleven miles from the next supply depot. Six months later, a search party found the team's frozen bodies in Scott's tent. The last diary entry read: "We shall stick it out to the end, but we are getting weaker, and the end cannot be far. It seems a pity but I do not think I can write more. R. Scott. Last entry. For God's sake look after our people." Praised as a tragic hero at the time, Scott has more recently been criticized for his reckless zeal to reach the pole at any cost, squandering the lives of his men.

These were not the last Antarctic expeditions. In 1914, Ernest Shackleton returned to undertake the first crossing of the continent across the pole, but before he even reached Antarctica, his ship, *Endurance*, became trapped in the ice. *Endurance* was carried along for a year before it was eventually crushed and split asunder. With supplies running low and no way to get home, Shackleton led a march across the ice, hauling small boats to ice floes that slowly drifted north. As the ice floes fragmented, the crew launched their boats and set out into the sea. They managed to reach Elephant Island, but this was an inhospitable rock, many miles from passing ships.* They'd already survived a year and a half, but now their only chance of rescue was an 850-mile open-boat journey to the whaling station on South Georgia Island.

Shackleton set out with five companions to bring help for the rest, sailing for weeks through stormy seas in constant peril of capsizing. As they reached the island, they were forced to ride out hurricane-force winds to avoid being dashed against rocks.† Even once they scrambled ashore, Shackleton and a single companion had to trudge over thirty-two miles of jagged rocks to reach the whaling station on the opposite side. Although he failed to achieve the goal he set when he first embarked on the expedition, Shackleton, by prevailing against so many obstacles to keep his crew alive and summon help, surely accomplished one of the greatest rescues ever. Thanks to his boundless resolve and unflagging leadership, not a single man was lost. A testament to his crew's loyalty was the willingness of many to sign up for his 1921 Antarctic expedition. In spite of the danger, or perhaps because of it, they couldn't resist the urge to explore.

* Thankfully, there were seals for food. Starved of carbohydrates, but with bellies full of meat and blubber, the crew read from cookbooks every night to rekindle memories of the tastes of home.

† The same storm sank a five-hundred-ton steamer sailing from Buenos Aires.

Antarctica was eventually conquered by technology. In 1929, Admiral Richard Byrd overflew the South Pole as he had the North three years before. Exploration became a scramble for ownership, as various contradictory claims were staked—the most sinister in 1939, when the German Antarctic Expedition flew over the continent, dropping tiny swastika flags across the ice. To forestall conflict, the Antarctica Treaty was signed in 1959, banning weapons and declaring that the continent should remain freely accessible to all nations for scientific purposes. The only continent without humans thus became the only continent without war. This international stewardship is likely to set a precedent for off-Earth claims on other worlds, and in many ways Antarctica is not so different from Mars's frozen tundra or Jupiter's moon Europa. Recent surveys have revealed lakes of liquid water buried deep under massive sheets of ice. Even in these isolated and inhospitable environments, we find life. This may be the most profound lesson we can learn from the remote regions of our planet: life strives to overcome challenges, and frequently thrives.

17 | TO THE SKY

It's truly remarkable that creatures who, until recently in evolutionary history, never traveled faster than a running pace or built more than rudimentary tools are today able to operate motor vehicles and create networks that span the globe. We developed the technology to design, build, and operate airplanes over a century ago, but there's still something mysterious, even unnatural, about them. We're so used to flying that we don't often think about how incredible it is that platforms of metal, supported by nothing more than curved protrusions directing airflow, can carry over a million pounds of structure, fuel, passengers, and luggage.* When we do ponder it, those of us who lack a close acquaintance with the principles of differential pressure can become nervous. But despite our apprehension, air travel is remarkably safe.

If you want proof of this, just look at the long lines of planes waiting to land at any major airport. On average, there's less than one major accident per month around the world, out of more than three million commercial flights. You're more than eighty times more likely to be killed in a car crash than in an airline accident. Even on the worst airlines in the world, your chance of being killed is less than your chance of winning the lottery (one in two million)—and the best airlines are far safer. Part of the safety is inherent. Danger

* The largest aircraft, the Antonov An-225, can lift 640 tons—equivalent to 150 African elephants.

in transportation stems from the potential to impact objects, which causes our bodies to decelerate faster than they can tolerate. Flying involves far more separation from potential obstructions, thus more reaction time. Being higher actually makes you safer.

The fundamental physics of flying means that airplanes can glide over long distances. A Boeing 747 has a glide-slope ratio of 15:1, meaning that from cruising altitude it could glide for a hundred miles, even if all its engines fell off. And engines fail vanishingly rarely, once every several million flight hours. Most planes can fly on a single engine, and losing all engines on a multiengine aircraft happens, somewhere in the world, no more than once every few years. Part of this reliability is due to extraordinary tests we subject engines to. We run them through temperature extremes; dump torrents of water, ice, and snow into them at up to eight hundred gallons per minute; and fire debris (including frozen chickens) at them with cannons. They're tough. Frequent inspections catch mechanical issues long before they become a problem, and even if one is missed, aircraft have redundant systems that allow them to fly with significant damage.

Additional safety comes from the design of airplanes, which can safely withstand G-forces that would terrify passengers. It's virtually impossible for an airplane to be damaged by turbulence: in fact, turbulence has never resulted in an airliner crash.* Engineers calculate what forces a structure should tolerate by assuming the absolute worst-case conditions and then pile safety factors on top of that. The size of these factors varies by structure: bridges and buildings are typically twice as strong as they need to be, while airplane safety factors are 150 percent or higher. What this means is that every engineered machine or structure you've ever encountered is massively overdesigned. Spacecraft and rockets generally have lower safety factors, because they couldn't function with the inefficiencies introduced by such overdesign.

Aviation wasn't always so safe. Most aviation pioneers crashed

---

* In rare cases, crashes have occurred when sudden gusts caused pilots to momentarily lose control near buildings or mountains, but the plane wasn't brought down by midair turbulence.

at least once. In 1908, five years after the first flight, the first fatality occurred when Lieutenant Thomas Selfridge was killed as a passenger on Orville Wright's plane. All six of the army airplanes delivered by the Wright brothers crashed within a year of delivery, killing eleven military pilots. Louis Blériot crashed two airplanes before becoming the first pilot to fly across the English Channel in 1909. The first long-distance flight between Constantinople and Egypt disappeared without a trace. World War I aviators were daredevils before they started shooting at each other, and most were killed by accidents instead of enemy action. Thrown into combat with only seventeen hours' training, the average pilot survived a few weeks. They weren't even issued parachutes, lest they abandon their aircraft too easily.

As pioneers pushed the limits, flying gradually evolved from a risky proposition into the safest means of transport. Despite the mythology surrounding Charles Lindbergh, he was nowhere near the first person to fly the Atlantic—at least eighty-one people did this before him, some solo, beginning with the British aviators Alcock and Brown, who used a modified bomber in 1919. Lindbergh was merely the most *famous* to do so, largely because he was American, was good at publicity, and won the prize for a solo nonstop flight between New York and Paris. Airplanes were slower in those days, so transatlantic flights took a long time. Forget five-hour cruises in commercial jets: Lindbergh spent thirty-three hours at the controls, incontestably a grueling feat of endurance. One of the biggest dangers he and other long-distance pilots faced was falling asleep at the wheel.

The first aerial circumnavigation was completed three years before Lindbergh's 1927 flight, when four aircraft set off from California. They took the shortest possible route, avoiding the Pacific by flying from Alaska to Japan. Three of the four planes crashed along the way, and maintenance stops extended the ordeal to six months. The second aerial circumnavigation, in 1928, included the first transpacific flight. Charles Kingsford Smith's four-man crew crossed from California to Australia in three segments, the longest of which took thirty-four hours, during which they flew 3,200 miles between Hawaii and Fiji. Several other milestones would be reached over the next decade, including a marathon forty-one-hour nonstop flight between Japan and Washington State in 1931, but none would capture the public imagination like the achievements of celebrity aviatrix Amelia Earhart. More than anyone, Earhart popularized air travel for a generation of Americans. Her modest demeanor and tousled hair made her a perfect heroine for a media-conscious age.

The sixteenth woman in America to earn a pilot's license, Earhart was the first to fly across the Atlantic. She was the first woman to solo across the country from coast to coast, 2,500 miles in nineteen hours. She broke various speed and endurance records, even those set previously by men in a profession dominated by males. She established flying clubs, published popular books and articles, and toured the country on public speaking tours. A visiting professor at Purdue, she taught aeronautical engineering and acted as a career counselor for female science and engineering students. She became vice president of the National Aeronautic Association and helped found one of the first airlines. She shattered barriers, advancing not only air travel but the spectrum of achievement for all people across the globe.

Amelia Earhart had always been a daredevil. As a child, she loved climbing trees, shooting rifles, and building homemade roller coasters. In one incident, she affixed a slide to the roof of her family's toolshed, accelerating down the incline in a cart. Emerging with a torn dress, some bruises, and a sense of exhilaration, young Amelia proclaimed her first "flight" a huge success. In 1920, she got her first taste of real flying when famous air racer Frank Hawks took her up for a ride that would change her life. By the time she lifted from the ground, she knew she was made for flying. Working a variety of jobs from photographer to truck driver, she managed to save enough money to pay for flying lessons. The next year, she took a bus to an airfield near Long Beach, California, walked up to the veteran pilot Neta Snook, and asked if she'd teach her to fly.* Within a year, she was setting

---

* Neta Snook was a pioneer aviatrix in her own right. She befriended Earhart and in later years became resigned to being remembered mainly as her instructor. Her autobiography was titled *I Taught Amelia to Fly*.

records and attracting attention with her cropped hair and worn leather flying jacket.

In 1924, Earhart enrolled at Columbia to study medicine but soon ran out of money and was forced to abandon her studies. She tried to enroll at the Massachusetts Institute of Technology but couldn't afford the tuition, so instead she found a job as a teacher and social worker. All the while, she kept up her aviation interest, joining flying clubs and writing newspaper columns. Soon after Charles Lindbergh's transatlantic flight, Amy Phipps Guest (a multimillionaire aviation enthusiast) was looking for a woman with "the right image" to "fly the Atlantic" and contacted Earhart. She was to be copilot, mechanic, and navigator for a male pilot but happily agreed. When interviewed upon landing, Earhart modestly understated her role as "baggage, like a sack of potatoes . . . but maybe someday I'll try it alone." Despite her secondary role, she was invited by Calvin Coolidge to the White House and proclaimed "Queen of the Air" in newspapers.

Vaulted into the public eye, Earhart found that her celebrity was a means to finance her flying, so she started to promote women's fashions in newspapers and magazines. And soon she was recording achievements "untarnished" by flying as copilot. In 1928, she soloed across the United States, and then in 1932 she repeated the feat across the Atlantic. In 1929, she helped establish the first regular passenger service in the United States, between New York and Washington, DC.*

Equally active in the social sphere, Earhart fought for women's rights. After receiving six marriage proposals, she reluctantly agreed to wed author George Putnam but insisted on keeping their union an equal partnership. Not only did she not take his name, but Putnam frequently found himself called "Mr. Earhart." She worked on women's causes with First Lady Eleanor Roosevelt, whom Earhart took flying, helping her earn an aviator's permit. All the while, Earhart longed for a greater challenge—an aerial circumnavigation of the globe. This wouldn't be the first, but it

* This was Transcontinental Air Transport, which later became TWA, acquired by American Airlines in 2001.

would be the longest up to that time, sticking as closely as possible to an equatorial route. With financing from Purdue University, she retrofitted a twin-engine Lockheed Electra 10E, known as *Electra*, with extra fuel tanks and instruments, and hired experienced Pacific navigator Fred Noonan as copilot.

The first attempt ended badly. In March 1937, after crossing from California to Hawaii, the plane was damaged during take-off from Pearl Harbor and had to be shipped back to California for repairs. By June, Earhart and Noonan were ready to try again, and decided this time to fly east across the continent. Then, flying south from Miami, they stopped in South America, Africa, India, and Singapore, finally arriving in New Guinea by June 29 after covering 22,000 miles (75 percent of the total distance). The last leg would be the most dangerous, over the vast span of the Pacific. The plan was to head for Hawaii, making just one refueling stop at tiny Howland Island along the way. The coast guard ship *Itasca* was stationed at Howland to broadcast a radio beacon that would guide them in. *Electra* never arrived.

What happened? The exact events are a mystery, but it seems clear that Earhart and Noonan never located the signal from *Itasca* and eventually exhausted their fuel somewhere over the Pacific. One reason may have been a frequency mismatch between the aircraft's radio and antenna. Radio navigation was in its infancy, and the system had never been properly tested. Alternatively, contemporary photographs suggest that the aircraft's antenna may have been damaged during takeoff from New Guinea. As Earhart approached Howland, she sent out several radio transmissions that were received by *Itasca*. The first was routine, reporting on weather and progress, but soon Earhart was requesting that *Itasca* turn on its radio beacon. Already transmitting, *Itasca*'s crew started to panic as they realized their signal wasn't being received.

Earhart requested bearings to Howland, but *Itasca* lacked radar, so they had no way of discerning where she was. To make matters worse, *Itasca* couldn't send voice on her frequency, so while they could hear her, they could only respond in Morse code. Soon she reported reaching the expected position of the island but couldn't see it and was running low on fuel. *Itasca* lit boilers to create smoke to guide her in, but to no avail. Her last clear signal was a position report, followed by a few garbled transmissions. With contact lost, the search for Amelia Earhart began. The navy vectored in all nearby ships, including the battleship *Colorado* and aircraft carrier *Lexington*. It was, to that point, history's largest and most expensive search. Even two Japanese ships lent a hand. Sporadic radio signals were reported for days, leading to speculation that *Electra* had crash-landed, but with so many ships and aircraft blanketing the airwaves, it's unclear if any were from Earhart. For months, vessels scoured Pacific atolls, looking for signs of the lost aviators.

It's possible they simply ran out of fuel and crashed at sea, but the leading theory has the pair crash-landing on remote Gardner Island, possibly injured, and surviving for days. This idea is supported by a number of clues, such as radio transmissions received after the crash and a 1937 aerial photograph showing what could be wreckage. In 1940, a British pilot found bones on the island, supposedly matching those of Earhart, but these later disappeared, compounding the mystery. Several suspicious artifacts have also been found, including a sextant box and folding knife, although it's difficult to disentangle these from the island's Second World War radio station.* One imaginative but entirely unsubstantiated proposal was the surprisingly widespread belief that Earhart was captured by the Japanese. Since *Electra*'s crash was only a few years before the war, a story was advanced that the aircraft was disassembled to inform Japanese aviation technology, and Earhart was either executed or taken to Japan as part of the "Tokyo Rose" program.† (Needless to say, this seems far-fetched.)

Soon after Amelia Earhart's fateful flight, the Second World

---

* The fact that there isn't enough evidence to confirm whether a modern aircraft crashed at Gardner Island less than a century ago illustrates the difficulty of confirming early contacts between the Old and New Worlds (chapter 7). Archaeological evidence must be substantial to survive for centuries.

† During the war, the Japanese broadcast messages to taunt Allied servicemen using English-speaking women collectively called "Tokyo Rose." This was intended to demoralize them but was actually a huge hit with Americans, who eagerly tuned in.

War inaugurated a revolution in aviation technology. It was the dawn of the jet age, and soon aircraft were flying higher, faster, and farther than ever before. Nowadays, one hundred thousand commercial flights take off every day, the largest carrying a thousand passengers (the record is 1,088 passengers, including two born on board, during a 1991 evacuation of Addis Ababa). Airplanes are safer than ever, thanks to the incremental efforts of the first trailblazers. Nevertheless, it's still possible to reflect on the wonder of flying, as early aviators once did. In June 2016, a solar-powered airplane completed the first circumnavigation of Earth without consuming a drop of fuel. Advanced environmentally friendly designs may soon transport cargo and passengers to far-off destinations. Earth-to-Earth rocket transport is also on the horizon, offering the prospect of rapid suborbital travel anywhere in the world within an hour. The future lies toward the sky.

18 | THE SPACE RACE

Going to space seems like a natural extension of flying—you just have to go higher. Air appears to be made of, well, nothing, and since the ancient philosophers, we thought it was just that. You can't swim through air as you can through water.* However, on deeper reflection, the notion that air is nothing is actually rather strange. Anyone walking outside in a breeze—let alone in a hurricane—surely notices that air can exert significant force. In fact, air only seems like nothing because it usually pushes on us with the same force in all directions. Air is heavy, forming a column rising to space of 14.7 pounds per square inch, and this means it can be pushed against. Airplanes and helicopters generate a pressure differential that lifts them up, like a breeze from below. Balloons float because they're lighter than air. Space is not quite nothing, but it *is* a quintillion times more sparse than air (1 followed by eighteen zeroes), with only a few atoms per square inch. If you want to travel in space, you can't push against air.

An early solution was proposed in Jules Verne's 1865 novel *From the Earth to the Moon*, where the crew was fired into space with a giant cannon. While this could theoretically work, such a cannon would have to be larger than the Empire State Building, subjecting the crew to around twenty-two thousand Gs: enough

* On the Space Station, you can in fact "swim" through air with your arms and legs, albeit slowly.

not only to kill them but to shred them at the cellular level. Clearly, there was work to be done. The problem was eventually solved in the least expected place: a log house in rural Russia, where an eccentric recluse was pondering methods of space travel. Between 1880 and 1935, Konstantin Tsiolkovsky wrote more than four hundred works on the subject. Although he never built rockets of his own, he developed the fundamentals: multistage boosters, maneuvering thrusters, bipropellant and cryogenic rocket fuels, regenerative cooling, space stations, airlocks, life-support systems, and interplanetary trajectories. His 1903 masterpiece, *Exploration of Outer Space by Means of Rocket Devices*, offered a complete blueprint for space travel, laying the groundwork for everything to come.

One of the first to test Tsiolkovsky's ideas was Robert Goddard, who in 1926 succeeded in launching the world's first liquid-propelled rocket. Building a test-firing stand, he experimented with nozzle shapes by measuring their thrust and exhaust velocities (the most important characteristics of rockets, which determine, respectively, power and efficiency). In engineering, "know-how" often precedes "know-why." You don't need to understand how something works to use it, so it's sometimes expedient to simply test lots of alternatives. This is what Goddard did. Eventually, he managed to build an engine with an exhaust velocity of seven thousand feet per second—low by today's standards, but nevertheless a major breakthrough that inspired all future rockets. Goddard had many disciples around the world, and one of these was the man who, more than any other, brought about the space age.

Wernher Magnus Maximilian Freiherr von Braun is an inspiring and controversial figure who first developed rockets as weapons for the Nazi regime and then moved to the United States to lead the Cold War race to the Moon. Von Braun started young: at age twelve he was arrested for strapping rockets to his wagon, which raced through town belching smoke and fire. Reading about Goddard's experiments, he harbored an insatiable passion to propel humans into space by any means necessary. Graduating from the Technical University of Berlin in 1932, where he worked under rocket pioneer Hermann Oberth, von Braun soon found himself living in the shadow of a militarizing regime.*

Von Braun wanted to go to the stars, but he needed funding to get there. The German military was interested in funding rocket research to circumvent Treaty of Versailles arms limitations (rockets weren't mentioned in the treaty). Von Braun's 1934 graduate thesis on liquid rocket propulsion was kept as a classified secret by the German army. Soon, the young man rose to lead the German rocket program, launching the first rocket to reach space in 1942. Serving the Nazis until the end of the war, he had a complicated relationship with the regime. He needed its support but considered the relationship a means to an end.† Although he enjoyed favor at the highest levels of Nazi leadership, he was also considered a flight risk and was once arrested by the Gestapo for questioning. As the war drew to a close, with Allied and Soviet forces converging on the heart of Germany, von Braun found himself at the top of a US list of German scientists to grab before they fell into Soviet hands. Von Braun and his rocket team surrendered to US forces on May 2, 1945.

Von Braun and his colleagues were transferred to the United States to work for the US Army, where they were issued false employment papers that wiped clean their Nazi past. The Cold War was on, and the United States now needed to focus on outcompeting the Soviet Union. The German rocket team's first job was to rebuild and launch V-2 rockets that had been captured at the end of the war and shipped to White Sands, New Mexico. This was the start of the American space program. In 1950, von Braun and his team were transferred to Redstone Arsenal in Huntsville, Alabama, which would eventually spawn the Marshall Space Flight Center, still home to NASA's rocket design programs. Working for the army in Huntsville, the German scientists cranked out a series of increasingly sophisticated rockets.

* Oberth was a father of rocketry in his own right, with his foundational 1922 and 1929 books *Die Rakete zu den Planetenräumen* (*The Rocket into Planetary Space*) and *Wege zur Raumschifffahrt* (*Ways to Space Flight*).

† The 1960 film of Von Braun's life, *I Aim at the Stars* is sometimes unofficially given the subtitle *But Sometimes I Hit London*.

Meanwhile, the Soviets were working hard to produce their own rockets. There was additional pressure on the Soviet side, due to US air superiority and European bases' being located closer to the Russian heartland. From the Soviet perspective, the only way to counter America's advantage of geographic remoteness was to develop a strike capability that couldn't be intercepted. This meant missiles that could fly through space to hit the United States directly. The race was on. Leading Soviet efforts was Sergei Korolev, who, like von Braun, was interested in peaceful space exploration but could find only military funding. With the help of a few German scientists, the Soviets won the race in 1957 with R-7, the world's first intercontinental ballistic missile (ICBM).* Korolev realized that the capability to strike the United States also offered the potential to deploy a peaceful satellite in orbit. Thus Sputnik was conceived, largely as a publicity stunt, but one that would change the world.

Sputnik's 1957 flight changed everything. A Soviet satellite circling overhead, emitting persistent radio chirps, simultaneously brought home to an anxious American public both potential dangers from the sky and the magnificent possibility of space travel. A month later the Soviets launched Sputnik 2, which carried the first animal into orbit—a dog named Laika. The space race seemed over, with the Soviets victorious. In response to Sputnik, President Eisenhower called for the creation of NASA, explicitly conceiving it as separate from the military and devoted to "peaceful purposes for the benefit of all mankind." Peaceful though its purpose may have been, NASA was intended to win the Cold War on at least the propaganda front. After some rushing to turn blueprints into workable systems, America was ready to launch its first satellite in December 1957, just two months after Sputnik, but disaster struck when the navy-built Vanguard rocket exploded on the launchpad. Von Braun's team of Germans was called in to save the day, and two months later their Jupiter rocket lofted the first US satellite into orbit.

* R-7 is a prolific rocket family: its variant Soyuz now launches astronauts (including American ones) to the International Space Station.

The United States had—just barely—caught up. It didn't last long. Russian rockets were more powerful than anything the United States had developed, so success followed success. On April 12, 1961, with the flight of Yuri Gagarin, the Soviets put a human into space for the first time.* It seemed Americans had learned nothing from Sputnik, and the Soviets had again won the race. Less than a month later, on May 5, Alan Shepard blasted into space, but his flight was a mere fifteen-minute suborbital stunt over the Atlantic, compared with Gagarin's orbit of Earth. Across the world, communism was starting to look pretty formidable, and the United States needed to do something dramatic. On May 25, 1961, President John F. Kennedy held a joint special session of Congress, where he proposed that the United States should "commit itself to land a man on the moon and return him safely to the Earth" before the end of the decade. Never mind that this was a political stunt to distract from the disastrous Bay of Pigs invasion of Fidel Castro's Cuba; America was embarking on history's greatest space endeavor, and the people were behind it. Kennedy's assassination two years later merely cemented this goal. If the United States had won the opening phases of the space race, it might never have gone to the Moon. But America lost, and now it had to catch up, redefine the goal, and then win.

The Soviets were still ahead, and would be for some time.† The Moon was carefully selected as a goal because it would also pose a very difficult challenge for the Soviets. This meant catch-up time. It's tough to express just how far the United States was from the Moon in 1961. Its first astronaut had just flown to the edge of

---

* Gagarin never touched the controls, which were locked except in case of emergency. While the Soviets equated mechanization with modernity, the United States emphasized "heroic pioneers." When the Soviets flew the first female astronaut, Valentina Tereshkova, in 1963, some Americans presented this as evidence that Soviet spacecraft were so mechanized that even women could fly them—which seems not only shockingly sexist but also odd considering that spacecraft on both sides had previously been "piloted" by a variety of dogs, monkeys, and chimpanzees.

† A selection of Soviet "firsts": first satellite, first human in space, first woman in space, first space walk, first lunar probe, first photo of the Moon's far side, first (robotic) landing on the Moon, first probe to another planet (Venus).

space for a few minutes. No one knew if humans could survive in space for long periods of time, how they could navigate to the Moon, how they could land there, or how they could step outside their spacecraft for a look. The Gemini program (1965–1966) was improvised as a way to learn how to do all these things. The flights were arranged in order of increasing difficulty, focusing on space walks, long-duration flights, orbital rendezvous, and docking maneuvers. By 1967, NASA had established that spacecraft could meet in orbit and that humans could, without going crazy, survive for two weeks in a capsule smaller than a Volkswagen Beetle.

Meanwhile, von Braun and his rocket team prepared their masterpiece, the most powerful rocket ever. With 7.9 million pounds of thrust, Saturn V was capable of lifting into orbit one hundred fifty tons—around twenty-seven African elephants. Von Braun initially argued for using multiple launches to assemble a large spacecraft in orbit that could land on the Moon, but by 1962 he came around to the idea of a "lunar orbit rendezvous." This plan involved a single launch to send a small craft to the Moon with an even smaller landing craft to make the final descent to the lunar surface. In fact, four spacecraft segments went to the Moon, and of the command module, service module, lunar module descent stage, and lunar module ascent stage, only the first—less than 0.2 percent of the total launch mass—returned to Earth. The Saturn V stack was 20 percent taller than the Statue of Liberty, but the crew made the entire journey in a ship the size of a minivan.

The first Apollo flights were intended to be test missions, but by late 1968 there was a heightened sense of urgency. With only a year left to meet Kennedy's goal, lunar module development was dragging on. Meanwhile, the Soviets were making significant progress and might at any moment have won the race.* With the deadline looming, NASA made perhaps the boldest decision in its history: it would reverse the order of the Apollo 8 and 9 flights, launching Apollo 8 to the Moon without the lander. This would be

* The Soviets had two separate Moon programs: one orbit, and one landing. Both programs were plagued by test failures, but if they'd concentrated on just one, they might have succeeded. After losing the Moon race, the Soviets downplayed the contest, claiming they'd never tried.

only the second Apollo launch with humans, and they'd be going on a weeklong journey around the Moon with no hope of rescue. Throughout 1968, America had seemed to be unraveling, with the North Korean capture of an American warship, the Vietnamese Tet Offensive, and the assassinations of Martin Luther King Jr. and Robert Kennedy. Then, on Christmas Eve 1968, Frank Borman, James Lovell, and William Anders became the first humans to gaze upon our planet from afar. *Earthrise*, taken as they circled the Moon, became perhaps the most influential photograph of all time. To look at the image was to grasp the fragility of humanity's lonely blue oasis set against the black void of space. As an anonymous observer put it to Borman, "Apollo 8 saved 1968."

A few months later, on July 20, 1969, an estimated 530 million people raptly stared at televisions across the globe as Apollo 11 mission commander Neil Armstrong stepped off the lunar module as the first human to set foot on another world. Between 1969 and 1972, six Apollo missions landed on the Moon, carrying a total of twelve astronauts to its surface.* The last three missions brought a rover to extend the crew's range, and the last featured a geologist for the first time: Harrison Schmitt, the only scientist to visit another world. Even so, Apollo barely scratched the surface. All landings were at equatorial sites on the near side, with a cumulative total of twenty-four hours of surface exploration across all missions. Saying we've "explored" the Moon would be like landing in six shopping mall parking lots and claiming to have explored Earth.

Ultimately, NASA was a victim of its own success. With the goal of beating the Soviets having been accomplished, public interest waned. Even as President Nixon was parading with Armstrong, Aldrin, and Collins, he was canceling the last three planned Apollo missions. NASA's budget was slashed by 80 percent, forcing tough choices. The plan had been to follow up with ambitious programs including a permanent lunar base, an Earth-orbiting space station,

---

* Apollo 13 was supposed to land until the explosion. Thus, Jim Lovell (played by Tom Hanks in *Apollo 13*) went to the Moon twice (Apollos 8 and 13) but never landed.

flyby missions to asteroids and Venus, a reusable space plane, and a mission to Mars. All were scheduled to be well under way by the 1980s, with a Mars landing planned before 1990. Facing impending budget cuts, NASA decided to focus on just one of these. The space plane was selected because it offered the prospect of reducing the cost of access to space, which would pave the way for more ambitious projects down the road. The resulting Space Shuttle was a technological marvel, but owing to design compromises, it became a massive strategic failure. Instead of reducing the cost of space access, it *increased* it. Unable to travel beyond Earth orbit, the shuttle was used for essentially the only mission for which it was suited: building a space station. Thus the world ended up with the International Space Station, not necessarily because it was the best choice, but because it was something we could do. The task of visiting other worlds was left to robots and future generations of explorers.

19 | SEEING THROUGH ROBOTIC EYES

A tiny robot named Sputnik inaugurated the space race, sending out a steady heartbeat of radio pulses.* Overshadowed in the public imagination by the human dimension, a parallel Cold War contest—between robotic explorers of competing ideologies—played out simultaneously. The advantage of using robots to explore space is that they don't need food, water, or air, so we can send them virtually anywhere with no plan for their return. Because space is a vacuum, space probes don't even need engines to continue their journey (though many use engines to maneuver or land). Some spacecraft, such as the Saturn-exploring Cassini, deploy rovers or smaller modules for focused missions, such as the Huygens lander that descended from Cassini into the thick atmosphere of Titan, Saturn's largest moon. A few return to Earth, bringing home samples from another world.

In 1959, less than two years after Sputnik, the Soviets sent the first robot to another world. Not bothering to figure out how Luna 1 could enter the Moon's orbit, scientists were content for the probe to smash into the lunar surface at six miles per second. However, Luna 1 missed the Moon (a surprisingly small target at that distance) and hurtled into permanent orbit around the Sun. Instead of a quick death, Luna 1 was granted virtual immortality.

---

* "Sputnik" = "Companion Traveler." The pulses served no purpose beyond signaling location.

Subsequently, Luna 2 succeeded in slamming into the Moon, earning the posthumous distinction of being the first Earth explorer to "land" on another world.* A third probe, Luna 3 (1959), flew past the Moon, sending back photographs of the far side.† (In January 2019, China became the first country to land on the lunar far side, with the Chang'e-4 spacecraft.)

In 1961, the Soviet Venera 1 probe became the first human-built object sent to another planet. All that was known about Venus was that it was about the size of Earth, probably made of rock, and covered in thick clouds. For all we knew, there could have been a thriving civilization there enjoying the balmy Venusian weather. Unfortunately, radio contact was lost with Venera 1 on its final approach, and Venera 2 suffered a similar fate. Venera 3 managed to send back data, but transmissions abruptly ceased once the probe penetrated the cloud tops. The same thing happened with Venera 4. What was going on? Venus, it turned out, is blanketed by an atmosphere ninety times thicker than Earth's; it traps heat in a runaway greenhouse effect that elevates temperatures to 864 degrees Fahrenheit.‡ Venus is the solar system's pressure cooker, hot enough to melt lead. Its clouds of sulfuric acid aren't especially inviting either. Venus developed into something of an obsession for the Soviets, who sent sixteen missions there between 1961 and 1984. Later probes sported armored shells, and some managed to land and send back photographs—our only pictures beneath the clouds of Venus. The most successful, Venera 12, survived almost two hours on the harsh barren landscape.

The Soviets also tried to visit Mars but had even worse luck than they did with Venus. Between 1960 and 1964, a total of seven Soviet missions to Mars all resulted in failure. This allowed the

---

* Despite seeming like a strange way to explore, impactors provide useful data by taking measurements as they descend, and by kicking up debris whose pattern can be analyzed.

† The Moon is tidally locked so one side always faces us, but there is no "dark" side of the Moon since both sides are illuminated evenly by the Sun. However, because the Moon gyrates in orbit, we can see more than half from Earth: 59 percent, to be precise.

‡ This is actually how we learned about greenhouse effects. Similarly, studying Mars alerted us to the dangers of nuclear winter.

United States to claim credit for the first successful Mars exploration, when Mariner 4 flew past the Red Planet in July 1965 at a closest approach of six thousand miles. Historically, Mars missions have experienced a cursed track record, especially Russian ones: of their twenty-one, only two have been successful. The US track record is almost the reverse: five failures and seventeen successes. The first robots to land on Mars were America's two Viking missions in the mid-1970s. Each Viking spacecraft consisted of an orbiter and a lander. While the lander descended to the surface, the orbiter took photographs from above and acted as a communication relay. These missions revolutionized our understanding of Mars by providing high-resolution surface maps as well as important data related to the planet's chemistry, meteorology, seismology, and—potentially—biology.

Mars is a cold desert, but it has a lot of water locked up in ice. If all this ice were melted, it would provide enough water to cover the entire planet in an ocean between a hundred and a thousand feet deep, with more water underground. Although the low temperature and pressure can't support liquid water on the surface for long, in 2015 NASA confirmed evidence of liquid water seeping onto the surface in the form of salty brine. On Earth, everywhere there is water, there is life. We think that Mars was once warmer and supported oceans for billions of years—ten times as long as it took life to develop on Earth. We also know from missions such as the Kepler Space Telescope that planets are extremely common, with billions capable of supporting life in our galaxy alone. So a question arises: Do worlds capable of supporting life develop life, or is life a rare phenomenon? The discovery of even long-extinct microbes on Mars would suggest that life is abundant throughout the universe and that we're probably not alone.

Although three of the experimental results gathered by the Viking landers were negative (suggesting a lack of organic molecules), the experiment for metabolic activity came back positive. Due to the lack of organics, this positive result has generally been dismissed as a chemical reaction, but there's another possibility. In 2008, the Phoenix lander found evidence of perchlorate salts, which destroy organics when heated (as the Viking samples were).

So the possibility that life was detected by the Viking landers remains open, because perchlorates may have destroyed organic molecules before they registered. Considering that Martian environmental changes took place over millions of years, that liquid water still exists underground, and that life thrives in the harshest environments on Earth, it seems likely that if life did develop on Mars, pockets persist to this day. In fact, pieces of Mars are occasionally blasted into space by impacts, to be found as meteorites on Earth. Since some bacteria can survive in space, and the two planets overlapped in habitability, life on Earth may have originated on Mars. We may be Martians.

Over the years, almost fifty robotic missions have been launched to Mars, around half successful. Progress has been steady but incremental. In six years, for example, the Curiosity rover, which moves *very* slowly, has traveled fewer than ten miles. At that rate, it would take seven thousand years to circle the planet, which would hardly mean visiting the entire place. In terrestrial terms, Curiosity would have done a decent job exploring New York's Central Park. With each mission, though, we discover something new about the planet and new intriguing questions form, such as: What caused Mars to lose its atmosphere and become a cold desert? And could such a thing happen to our planet, or to other planets out there, orbiting other stars?

While we've been exploring Mars, we've also been sending probes into the outer solar system. In 1972 and 1973, Pioneer 10 and 11 became the first spacecraft to take close-up photographs and measurements of Jupiter and Saturn. But this was merely the appetizer. In 1977, a pair of spacecraft, Voyagers 1 and 2, were sent on a "grand tour" of the solar system, taking advantage of a rare alignment of the outer planets. Voyager 1 followed a more direct trajectory, but Voyager 2's course enabled it to pass close to not only Jupiter and Saturn but Uranus and Neptune, too (so far, Voyager 2 is the only spacecraft to visit these "ice giants").* These missions produced most of the images of the outer planets we're

---

* Jupiter, Saturn, Uranus, and Neptune are giant balls of gas bound by gravity, but Uranus and Neptune contain higher proportions of ice.

familiar with and discovered most of these planets' moons.* After completing their primary missions, the Voyager spacecraft drifted out toward interstellar space. In 1990, Voyager 1 was given one last photographic assignment, turning its camera back toward the Sun to take a "family portrait" of all the planets in the solar system from afar. Out of 640,000 pixels composing the frame, Earth occupied less than one—truly a pale blue dot.

After forty years in space, the Voyager spacecraft are now traversing the depths between stars. Each carries directions to Earth, in a pulsar map that should be decipherable to extraterrestrials, no matter how different their language. Also aboard is a golden record carrying sounds of planet Earth, including greetings in fifty-five human languages and one whale (humpback, to be precise, although we have no idea what the whale was saying). It seems unlikely that either spacecraft will ever be picked up by another life form, but perhaps the message they carry is as much for us as for anyone else. The statement "We're here" is an important reminder of our place in the cosmos. And since the Voyagers will endure into the waning days of the universe, long after Earth has been consumed by the Sun, who knows what might come of them someday out there?

From 1989 to 1995, the Galileo spacecraft traveled to Jupiter, following slingshots around Venus and Earth.† It was appropriate that this first mission to Jupiter should be named after the man who gazed up with a telescope and observed that the planet was orbited by moons, proving that Earth was not the center of the universe. Along the way, Galileo passed an asteroid, Ida, and found that it, too, had a small moon, Dactyl. Approaching Jupiter in 1994, Galileo captured Comet Shoemaker-Levy 9's thunderous impact with the planet. The fireball slammed into Jupiter with more force than the comet that killed off all the dinosaurs on Earth. With ten thousand times the explosive power of the bomb

---

* At last count, moon tallies are: Jupiter, sixty-seven; Saturn, sixty-two; Uranus, twenty-seven; and Neptune, fourteen.

† When Galileo passed Earth, it conducted a search for life. Based on atmospheric methane and radio signals, life on Earth was considered probable but not conclusive.

dropped on Hiroshima, Shoemaker-Levy 9 left scars on Jupiter's surface larger than Earth itself. (So, no dinosaurs on Jupiter then.)

Entering orbit the next year, Galileo surveyed Jupiter's moons, confirming a liquid ocean under the ice sheets of Europa that contains more water than all the oceans of our planet. Galileo also sampled volcanic plumes from Io, which belch sulfur dioxide snow hundreds of miles into space, and found that Jupiter, too, has a ring, much fainter than Saturn's. After eight years in orbit, Galileo was intentionally crashed into Jupiter in 2003 to avoid contaminating the moons with microorganisms that may have hitched a ride from Earth.

The year after Galileo's death plunge, the Cassini spacecraft arrived to study the Saturn system, sending the detachable Huygens lander to explore Saturn's moon Titan. For ninety minutes, Huygens transmitted data as it descended into Titan's clouds and touched down on the moon's surface—the most distant place we've ever landed. Along the way, Huygens mapped the moon's lakes,* and photographed the alien but somehow eerily Earth-like surface. Like Voyager, Cassini took breathtaking snapshots of our pale blue dot from afar, reminding us of how precious our tiny speck of a planet truly is. Cassini's surveys of Saturn's rings changed how we think about planetary formation; the rings, it turns out, play a role similar to the swirling dust clouds of young solar systems. Some moons, such as Titan, were already known to be interesting, but the greatest surprise was that even tiny Enceladus, smaller than the island of Britain, contains a subsurface ocean, adding it to the short list of places in our solar system able to support life.

Exploration of the outer solar system continues apace. In 2016, the Juno spacecraft entered Jupiter orbit to pick up where Galileo left off, measuring the planet's composition and magnetic fields. And NASA is now planning a mission to Europa, which would be only the second mission ever launched (after Viking) with a primary goal of searching for life on another world. This "Europa

---

* Titan is the only other world with surface liquid, and the only one other than Venus and Earth with a thick atmosphere.

Clipper" spacecraft will not only take measurements from orbit but actually descend to land on Europa and sample its icy surface.

The most numerous bodies in the solar system are asteroids and comets, chunks of ice and rock ranging from specks of dust to the dwarf planet Ceres, a quarter the size of the Moon. In 2006, the Stardust spacecraft returned to Earth with samples from the dust trail of a comet, capturing grains in aerogel as if collecting bullets in a pool of Jell-O. A few years later, the Japanese Hayabusa ("Peregrine Falcon") spacecraft managed to return to Earth samples from the surface of asteroid Itokawa. These samples represent the only materials so far returned to Earth from any world beyond the Moon (a few similar missions are currently underway). In 2007, the Dawn spacecraft was sent to study Vesta and Ceres, the largest bodies in the asteroid belt between Mars and Jupiter. But perhaps the most celebrated mission was Rosetta, which in 2014 orbited comet 67P/Churyumov-Gerasimenko and deployed the tiny Philae lander to its surface. All these activities are critical, not only for understanding the origins of the solar system but for practicing rendezvous and landings on asteroids and comets. Knowing how to operate on the surface of these objects could enable us to one day divert a potentially deadly threat like the one that killed the dinosaurs.

Our probes have now explored the outer reaches of our solar system, starting with the 2015 New Horizons flyby of Pluto after a nine-year voyage. Out at Pluto, the Sun provides less heat than a full Moon on Earth. On New Year's Day 2019, the probe took photographs of the most distant object ever explored, Ultima Thule, named for the Arctic observations of none other than Pytheas the Greek, who set out into the unknown over two thousand years ago. New Horizons is bound for the infinite void of interstellar space, but its mission will continue until at least 2021 as it observes other small icy worlds in the Kuiper Belt.

Someday it may be possible to send robots to other stars. But for now, we can use telescopes to peer out into the darkness. Space-based telescopes, which are unhindered by Earth's light and atmosphere, have been particularly effective, scanning every band of the electromagnetic spectrum from microwaves to gamma rays. Most celebrated is the Hubble Space Telescope, launched in 1990 and still going strong. Surveying the distant universe back to its birth,* Hubble is often considered the most successful space exploration mission ever, in terms of both science and public engagement.

In the last few decades, we've developed techniques to peer beyond our solar system to planets orbiting other stars (called "exoplanets"). The most successful mission to do this is the Kepler Space Telescope, launched in 2009. It has the ability to stare at 145,000 stars simultaneously, watching for vanishingly small dips in light as planets pass in front of them. It's a bit like standing in New York and keeping an eye out for flies passing in front of searchlights in Paris. With sensitive instruments, Kepler is capable of not only finding planets but discerning a lot about them, including their mass, size, composition, and distance from their sun. By this means, thousands of planets have been discovered, many Earthlike. It seems our blue marble is no aberration: there are more Earthlike planets in our galaxy than humans in existence, and more such planets in the universe than grains of sand on all the beaches of Earth. It's a big universe out there, waiting for us.

---

* Looking into space is peering back in time, since light takes substantial time to reach us. If Andromeda aliens could see us with a giant telescope (two million light-years away), they'd be observing the origins of humanity two million years ago. The most distant objects we see are billions of light-years away, from the infant universe immediately following the Big Bang.

PART IV

BECOMING *STAR TREK*

20 | INTO THE FUTURE

Earth has existed for about 4.6 billion years (a third the age of the universe), and it didn't take long for life to appear. Our planet has been inhabited by single-celled organisms for 80 percent of its existence, but multicellular creatures are another matter. Primitive plants and animals only appeared within the last 13 percent of Earth's existence (600 million years ago), and it took another couple hundred million years for anything larger than a microbe to venture from the sea onto dry land. Mammals appeared in the last 250 million years, primates in the past 50 million, apes in the past 13 million, and humans within the last million. If Earth's history were a day, life appeared at 4:00 a.m., fish at 10:24 p.m., dinosaurs at 11:15 p.m., and humans at 11:59 p.m. All recorded history took place within the last quarter second, and Columbus sailed at a hundredth of a second to midnight.

Not merely biology but technology has followed this accelerating trend. The first advances—stone tools, fire, animal-skin clothing—took hundreds of thousands of years to develop. Agriculture, writing, and the rule of law took many thousands. For people living even a millennium ago, technological change was virtually unnoticeable within a lifetime, and some centuries actually saw regress. It's been less than two hundred years since we advanced beyond animal muscle power and sailing ships. With the advent of steamships, hazardous multiyear journeys around the world were cut to a few weeks. Railroads reduced transcontinental treks to a few days of leisurely travel. Telegraphs and undersea cables allowed messages to race ahead of people, sprinting around the world in less than a day.

In 1899, the commissioner of the US patent office is supposed to have remarked that "everything that can be invented has already been invented." While this quote is probably apocryphal, humans have always had a tendency to think this way. We acclimatize easily. We're precisely the same biological organisms as we were thousands of years ago, equally at home hunting gazelles with sharpened sticks as programming supercomputers. It's easy to look around and see what we have, but it takes ingenuity to imagine what will be. Yet, technological development is all around us. Many of us grew up without personal computers, let alone smartphones harnessing more computing power than the spacecraft that took us to the Moon. Social media is only a decade old. Only the most devoted computer nerds had heard of the Internet thirty years ago.

In 1965, Gordon Moore, a founder of the microchip maker Intel, observed that the number of transistors on a computer chip doubled every couple of years.* This so-called Moore's law has held true ever since, doubling computing power every two years for half a century. In 1971, Intel's computer chips used over two thousand transistors, each costing a dollar. Forty years later, computer chips packed in seven billion transistors, each costing a hundred-thousandth of a penny. This rapid rate of progress, representing improvement by a factor of a million, suggests that, technologically, the next century will be more like ten thousand years than a hundred. Indeed, computing power is increasing so fast that some experts, such as Ray Kurzweil, believe we're approaching a "singularity" in which runaway exponential growth will spawn an artificial superintelligence.

Automation has dramatically reduced the need for human labor, increasing productivity and efficiency, but creating more with less has always exacted a social cost. Workers' fear of losing their jobs to machines dates back to the industrial revolution. But automation hasn't so much resulted in an outright loss of jobs

* Transistors are tiny switches that can be triggered by electric signals.

as driven a shift to more skilled sectors. New technologies dramatically alter the playing field, eliminating some industries while spawning others. While there's no guarantee that this equilibrium will hold forever, an automated future is a productive one, potentially enabling humans to pursue their unbounded aspirations. As *Star Trek*'s Captain Jean-Luc Picard explains from the twenty-fourth century, "A lot has changed in three hundred years. People are no longer obsessed with the accumulation of 'things.' We have eliminated hunger, want, the need for possessions. We work to better ourselves and the rest of humanity."

It's hard to imagine what benefits might be unlocked with artificial intelligence, which would approach problems with not just immense power but a different perspective from our own. We could be on the brink of ending disease, hunger, and poverty— but the stakes are high. Even if an artificial intelligence bears no explicit malice toward us, the unforeseen consequences could be disastrous. As suggested by Elon Musk, an artificial intelligence that controls a hedge fund might maximize profits by shorting consumer stocks, buying defense stocks, and starting a war. As with the proverbial monkey's paw, it's tough to prepare for all eventualities.* If you program a self-driving car never to run a red light, it might try to hack the traffic system to change the light to green. Like ants happily swarming over an anthill, we might get bulldozed simply for being in the way. Lacking emotions or human contexts, computers are natural sociopaths.

The rise of artificial intelligence is just one of the challenges we'll have to navigate if we want to survive to see a *Star Trek* future. From malicious robots to climate change, nuclear war, deadly viruses, asteroid impacts, nearby supernovas, supervolcanoes, and alien attacks,† the list of potential civilization-ending threats is getting longer, and there are likely other threats of which we aren't yet aware.

But we live at a unique time. For the first time in history, a spe-

---

* In a 1902 story by W. W. Jacobs, a monkey's paw grants three wishes, which come true, but in horribly unexpected ways.

† Signs of our existence in the form of radio signals have been leaking into space for a century, so if anyone nearby is listening, they'll soon know we're here.

cies has the technology to destroy itself in the blink of an eye. But also for the first time, and for precisely the same reason, a species is able to leave its planet and expand into space. After millions of years of evolution, we have arrived at the point where we can travel to other worlds. But since Earth is the most habitable planet we know of, why should we bother?

Perhaps the simplest reason is basic survival. A sustainable off-Earth presence would provide an insurance policy of last resort against many of the threats we face. Considering how little this would cost,* isn't it simply negligent *not* to take this precaution against extinction? An off-Earth presence would not only insulate us from disaster but might help prevent it. It was by studying neighboring planets that we learned of perils to our own. Venus's sweltering blanket of carbon dioxide awakened us to the dangers of runaway greenhouse effects. Planetary dust storms on Mars forewarned a Cold War Earth of the potential of nuclear winter, providing one more reason to step back from the brink of mutual assured destruction. If we're contemplating action on climate change by geoengineering Earth's natural systems, wouldn't it be nice to have a few comparative planetary case studies first?

Sustaining humanity on Earth is of course far more important than establishing it on other worlds. However, far from being in competition with this goal, a human presence in space would significantly advance it. Spaceflight is a technology driver because most breakthroughs come about indirectly. The entire field of antibiotics began when Alexander Fleming carelessly allowed fungus to grow on bacteria plates that should have been sterilized. Microwave ovens were invented after an engineer accidentally melted a chocolate bar using a radar set. Any investment at the leading edge is bound to produce spinoffs, but the multidisciplinary challenges of spaceflight are especially prone to drive innovation. A partial list of technologies inspired by spaceflight includes: air-traffic control, bomb-detection devices, bulletproof vests, camera phones, cordless tools, digital computers, fireproof clothing, insulin monitors, lap-

---

* Many people think NASA's budget rivals the military's, but it's actually only 2.5 percent as much (0.5 percent of the federal budget).

tops, life-support technologies, lightweight alloys, medical scanning technologies, remote control, scratch-resistant lenses, solar panels, telemedicine, water filtration systems, weather forecasting, and wireless switches.* The scientists scanning for black holes using the Hubble Space Telescope weren't intending to develop a means of diagnosing breast cancer, but, accidentally, this is precisely what they did when astronomical spectrographs proved just as effective at classifying tumors as black holes.

Since technology improves over time, can't we wait to travel to space? The short answer is no. We've made massive strides in computing since the 1960s, but little has advanced in spaceflight because we haven't been pushing our limits. Breakthroughs don't come along if we just sit around waiting for them. Technology follows purpose, not the other way around. Hyperadvanced propulsion, let alone warp drive, won't be invented unless we start by doing what we can with the technology we have. Columbus sailed to the Americas using flimsy coastal vessels because large ocean-going ships hadn't been invented yet—and never would have been without the incentives generated by the voyages. It is only by pushing the envelope of the possible that we guarantee advancement.

We face tremendous pressures from growing populations and diminishing resources, but scarcity is purely a matter of technology. Each hour, Earth is bombarded by enough solar energy to power our civilization for a year. Our world is swimming in water, so why do people go thirsty? Technology is the best—perhaps the only—way to sustainably raise the standard of living for everyone. The innovations required to live in space are precisely the same as those required to save our planet, but for space settlement there's a direct imperative. Easy to postpone on Earth, technologies such as portable water filtration, efficient small-scale food production, solar energy capture and storage, and on-demand 3-D printing are necessary for basic survival on other worlds.

As an exploring species, traveling to space gives us perspec-

---

* Contrary to widespread belief, NASA did not invent Tang, Velcro, or Teflon, although it did help popularize them. NASA has a full list of spinoff technologies on its website.

tive. How petty and insignificant our earthly squabbles seem from orbit. Spaceflight underscores the importance of preserving our planet, the only oasis we currently have. The environmental movement was fueled by this realization—in part thanks to the spectacular *Earthrise* photograph taken as Apollo 8 orbited the Moon in 1968. Collaborative enterprises beyond Earth bring humans together from across the globe. Imagine the social impacts of having settlers on another world. What unique perspectives would they develop? What problems would they solve? What would they be able to teach us?

Our most vibrant societies have always been outward looking. Exploration harnesses our restless energies toward constructive purposes. Life shouldn't only be about basic survival. Space travel offers the promise of a future worth living, a reason to get out of bed in the morning. The goal of settling other worlds would act as a call to our youth to develop their technological skills, so that they, too, might participate in the greatest adventure of our age. This is a challenge we've faced many times throughout history, one at the very heart of our nature. Once wanderers on the plains of Africa, we've now come full circle in our travels. We stand at the shore of a cosmic ocean, one that can never be tamed. From a world that has become too small, we're moving out into a universe that will forever be too large.

21 | THE ROAD TO MARS

Space is really close. If you could drive your car straight up, you'd reach it in under an hour. While distances on Earth's surface may be measured in thousands of miles, astronauts on the International Space Station regularly come within a few hundred miles of over 90 percent of all humans on the planet.* But the Space Station isn't even close to the lower limit of space, which lies just sixty-two miles above our heads. If you sat on a platform at the edge of space and jumped, you'd fall back down to Earth.† So why doesn't this happen to spacecraft? In fact, it does. If objects in space don't travel fast enough, their trajectory will cross our planet's atmosphere and they'll reenter in a fiery path as they compress the thin atmospheric air. The only reason anything stays in orbit is that it's traveling sideways really fast—so fast that by the time an object would fall back into the atmosphere, the atmosphere is no longer there. To stay in orbit, you have to travel ten times faster than a bullet: around five miles per second.

Getting into orbit is barely within our capability. The mathematics of the rocket equation dictate that even our best rockets can

---

* The Space Station orbits at an altitude of 250 miles, with an inclination of 51.6 degrees, meaning it will pass above everyone living between +/-51.6 degrees of latitude, a band in which 90 percent of the planet resides.

† In 2012, Felix Baumgartner jumped from a helium balloon twenty-four miles high (less than halfway to space). In doing so, he became the first human to break the sound barrier without an engine.

only lift a small percentage of their initial mass into orbit, and this relies on shedding most of the rocket along the way in the form of staging. It seems an astronomical coincidence that we live on the knife edge of being able to travel to space at all. If we lived on a planet like Mars, with only 38 percent of Earth's gravity, space travel would be easy. If we lived on a planet like Jupiter, with 253 percent of Earth's gravity, it would be nearly impossible.* Looking outside our solar system (specifically, 39.6 light-years from the Sun), the TRAPPIST-1 system has at least three planets in the habitable zone. Their lower gravity means that rockets launched from them can be half the size of ours. What's more, any inhabitants of these worlds, if they exist, will never want for destinations, with habitable planets almost as close as our Moon looming just as large in their sky.

Although humans have made great technological strides since 1969, it's not much easier to get to the Moon. Computers are smaller and faster, but they're a tiny fraction of a spacecraft's mass, and even less of a rocket's. In terms of actual preparation, we've regressed in our spacefaring abilities. Saturn V was the most powerful rocket in history, but it last launched in 1973. Since then, the human spaceflight program focused on the Space Shuttle and International Space Station, mainly because political and budgetary constraints haven't allowed for a bolder vision. No president or Congress has been willing to kill the American space program, but neither has any wanted to spend more on what is considered a luxury. Subsisting on only a half percent of the federal budget and suffering from a dearth of executive leadership, NASA has lingered on life support for half a century.

The emphasis placed on developing and operating the Space Shuttle is partly to blame for the lack of accomplishment. One of the shuttle's chief drawbacks was that its hybrid design combined cargo and crew so that every time you wanted to launch a satellite, you had to launch people, too. You certainly don't want to put people at risk for routine cargo runs, nor do you want to

---

* Of course, Jupiter is a gas giant with no surface. The largest stable rocky planet would be about ten times Earth's mass, or one thirtieth Jupiter's.

incur the extra cost of bringing unnecessary passengers along for the ride. Conversely, the requirement to carry large payloads burdened the crew vehicle with a massive superfluous cargo bay. Initially envisioned as the single launcher for all US space missions, the shuttle was driven to greater complexity by military requirements related to deploying surveillance satellites during the Cold War. But the military was never satisfied with the shuttle's capabilities or security, so it pulled out of the program, reverting to traditional rockets. The result was an overly complex vehicle that was too expensive to operate. In trying to do everything, the Space Shuttle ended up achieving next to nothing.

In stark contrast to the Saturn V Moon rocket, the Space Shuttle was the least efficient launch vehicle ever operated. There are various ways to tally cost, including cost per flight, total program cost, and average cost over a period of time. The Space Shuttle comes up short in all of them. In current dollars, a Saturn V cost just over a billion dollars per flight. Even adding a couple hundred million for assembly and launch, at 125 metric tons to low Earth orbit, this works out to less than $4,200 per pound to orbit. The cost per shuttle flight has been estimated at around $450 million. With a capacity to orbit of only 25 metric tons, this works out to almost double the cost of Saturn V, at something like $8,200 per pound. If we include development costs, the Space Shuttle should come out on top because the nonrecurring costs are spread over many more flights (135 versus 13 for Saturn V), but the shuttle was expensive to design as well as fly, so Saturn V comes out ahead by almost a factor of three!*

The shuttle program did perform some useful science and carried more than five hundred astronauts to space. However, accomplishment simply can't be measured in terms of humans carried into orbit—in fact, the higher number of launches actually had a negative impact on public opinion, making space travel seem routine and uninteresting. Fewer but more impactful missions would

---

* While it is true that Saturn V was just a launcher, and the shuttle could repair and retrieve satellites, this capability was rarely used—mainly because it was cheaper to simply build a replacement satellite instead!

have provided far more utility at a lower cost. The human space-flight program needs to have a coherent plan and specific destination in order to function efficiently. Fortunately, there seems to be a broad and growing consensus that it's time for the human space-flight program to leave low Earth orbit, and this time the destination should be Mars.

It's probably unrealistic to assume that going to Mars will be a repeat of the Moon race. If we made political decisions based on exploratory goals, NASA's budget would rival the military's and we'd have giant nuclear vessels plying not only the sea lanes but the space lanes as well. The reason Americans were willing to pay for the Moon was the Cold War's clash of ideologies. In Kennedy's speech to Congress in 1961, he made it clear—two paragraphs before the part about "landing a man on the Moon and returning him safely to the Earth"—that the entire program amounted to a public relations campaign to steal the Soviet spotlight.* After Kennedy's assassination, the program became part of his legacy, and it's unclear if robust funding would have continued had there not been near-unanimous support for seeing the fallen leader's vision fulfilled.

It's possible that another Apollo moment might come along. China is now firmly the world's second-largest economy and has lately been making steady strides in space. It has flown almost a dozen astronauts, built a space station, landed two rovers on the Moon (one on its far side), and has plans to build a permanent lunar base within the next twenty years. Russia, Europe, and Japan all have strong space programs. Even India is getting into the space race, sending a probe to Mars for less than it would cost to make a Hollywood movie about it,† with plans to launch its first astro-

---

* Relevant lines: "Finally, if we are to win the battle that is now going on around the world between freedom and tyranny, the dramatic achievements in space which occurred in recent weeks should have made clear to us all ... the impact of this adventure on the minds of men everywhere, who are attempting to make a determination of which road they should take."

† India was the first country to enter Mars orbit on the first attempt with its Mars Orbiter Mission, with a shockingly low price tag of $73 million, including launch.

naut within the next few years. The United States has historically shown not only a tendency for exceptional innovation but also a compulsion to be first and best at everything. How would Americans react if a rival nation were on the cusp of surpassing their Moon landing with the first voyage to another planet?

In the long run, space needs to pay for itself. However, large government investments in new transportation technologies are the norm rather than the exception. Steamships, the railroad, and airplanes were invented privately but then developed to commercial viability by government support. As we've seen, Columbus's expeditions were government funded, as were the voyages of da Gama, Magellan, Hudson, Cook, and almost every explorer throughout history. Most overseas settlements were government ventures. The Spanish, Portuguese, English, and Dutch outposts established around the globe during the Age of Exploration were paid for by governments, entirely or in part. Even the ostensibly private Virginia, East India, and Hudson's Bay companies wouldn't have survived without royal protection and support.

A primary role of government is to promote long-term prosperity beyond the investment horizon of the private sector. However, relative immunity from financial accountability is precisely why governments also tend to be inefficient. Private industry responds to circumstances in a more agile way, with incentives to maximize results while minimizing costs. In an ideal world, the government builds the railroads, and private industry figures out how to profit from them. In the case of spaceflight, government support can help foster technologies that support a robust private sector.

Besides placing satellites in orbit (part of a $250 billion industry), private companies can find a source of revenue in space tourism. Between 2001 and 2009, seven wealthy individuals paid between twenty and forty million dollars to fly on a Russian rocket to the International Space Station. A company called Bigelow Aerospace has already built several prototype inflatable modules, including the Bigelow Expandable Activity Module (BEAM), which was launched to the Space Station in 2016. BEAM is intended to remain connected until at least 2020, during which time the module's leak rate, temperature, and radiation exposure will be monitored. Expanding on this, we can imagine a future of luxurious space hotels catering to the super wealthy. Such orbital destinations are well beyond the financial means of most Earthlings, but short suborbital hops into space, currently priced at a couple hundred thousand dollars, are much easier to achieve.* (You only have to be a millionaire, not a billionaire!)

The most successful private space company is SpaceX, established in 2002 by PayPal cofounder Elon Musk with the express purpose of developing transportation systems to reach Mars. SpaceX was the first private company to launch a liquid-fueled rocket into orbit (Falcon 1, 2008); successfully launch, orbit, and recover a spacecraft (Dragon, 2010); and send a cargo flight to the Space Station (2012). With its ability to make regular supply flights to the Space Station and return, SpaceX's Dragon capsule offers the only means of returning science cargo to Earth.† As of January 2019, SpaceX has flown more than sixty successful launches and soon plans to fly astronauts under a NASA commercial crew contract. In February 2018, SpaceX launched the world's largest operational rocket, Falcon Heavy, which blasted Elon's cherry-red Tesla roadster and its "Starman" driver on a path to Mars. Also aboard was a plaque with the names of all SpaceX employees, including mine.‡ (Instead of one booster of nine engines like Falcon 9, Falcon Heavy uses three boosters for twenty-seven total.)

SpaceX's most important achievement has been to reuse rockets for the first time in history. After most rockets consume their fuel, they free-fall back to Earth, breaking up in the atmosphere or upon

---

* Imagine firing cannonballs in successive arcs, higher and faster. Suborbital flights are like cannonballs that arc up into space and then come back down to Earth. The first US astronaut, Alan Shepard, flew a suborbital trajectory in 1961.

† The Russian Soyuz returns astronauts but not large cargo. Most other supply craft burn up in the atmosphere, usually filled with trash.

‡ Other Easter eggs aboard: a circuit board with the inscription "*Made on Earth by humans*," "Don't Panic!" displayed on the car's dashboard control panel (a reference to *The Hitchhiker's Guide to the Galaxy*), a miniature version of the car and Starman on the car's dash, a digital repository of Isaac Asimov's *Foundation* series, and David Bowie's "Life on Mars?" playing on the radio.

impact with the ground.* On initial consideration, this seems reasonable because the point of a rocket is to launch something into space, and you need a lot of fuel to do that. But how much fuel? Since a rocket is over 90 percent fuel by weight—essentially a flying fuel tank—an empty rocket is pretty light. And since rockets use multiple stages, once the upper stage(s) and payload† have separated to continue up to orbit, the first-stage "booster" that got them off the ground has nothing left to carry but itself. This means that an empty booster can actually return on a surprisingly small amount of fuel, far less than it took to get up to the edge of space. Landing a booster downrange on a ship at sea or returning it to a shore-based landing pad costs the rocket only a small fraction of its overall performance.

Since a rocket's first-stage booster represents about 80 percent of its cost, if you can save the booster, you've already realized substantial savings. However, it's technically feasible to return other components, such as upper stages and aerodynamic nose covers (called "fairings"), and this might be done in the future. SpaceX fairings now carry parachutes and are aimed to descend into a giant net borne by a boat dubbed *Mr. Steven*. As Elon puts it, if there were a pallet of $6 million floating down from the sky, wouldn't you try to catch it?

Piece by piece, SpaceX rockets are becoming more reusable. The vast majority of a rocket's cost comes from construction (fuel is less than 0.5 percent), so reusability is the clear way to make space less expensive.‡ The eventual goal is aircraftlike operations. Airplanes land, passengers disembark, there are a few routine inspections, the fuel tanks are filled, new passengers board, and you fly again. An airliner is comparable in price to a rocket. If we threw away airplanes after their first flight, imagine how expensive flying would be!

---

* Launch sites are located on the coast because rockets launch over ocean (or wilderness in the case of Russia and China).

† Payload = whatever is being launched, like a satellite, crew vehicle, or cargo ship.

‡ The Space Shuttle was partly reusable, but it also required copious maintenance. Additionally, the reusable orbiter was heavy, cutting into the system's payload capacity.

Reusability should create a virtuous cycle whereby flights become more common, more reliable, and less expensive all at the same time, as with the evolution of aviation from early daredevils to modern commercial flights. With reusable rockets, we can imagine a future not only of travel beyond Earth but of commercial rocket flights around the world. Using suborbital trajectories, a rocket should be able to transport passengers anywhere in the world in under an hour. Instead of embarking on a grueling fifteen-hour flight from New York to Shanghai, a rocket passenger could cover the same distance in thirty-nine minutes. Voyages from Europe to Australia used to take months. Soon they could take minutes. The longest part of the journey would be getting to the spaceport. In this and other ways, a future of reusable rockets could directly improve the lives of not only those who choose to travel to space but also those who decide to stay home.

Other companies are also aiming for reusability. In 2015, Blue Origin, founded by Jeff Bezos of Amazon, successfully launched its New Shepard rocket on a suborbital flight just across the edge of space (one hundred kilometers) and then landed it. New Shepard is aiming to carry tourists on short hops to the edge of space, but Blue Origin is now designing a new class of larger orbit-capable launch vehicles, also reusable. So there might be a space race of sorts after all. Meanwhile, NASA has its own crewed spacecraft (Orion) in final development and is completing the design of a massive new launch vehicle, the Space Launch System (capable of 7.2 million pounds of thrust, just shy of a Saturn V). Orion and the Space Launch System would give NASA the capability to send humans beyond orbit for the first time since 1972. Humans are thus poised to reach out across the ocean of space once again. But where should we go?

We live in a solar system with many possible destinations, but only a few make the short list. Sending humans on a decade-long trip to Pluto (each way) is clearly beyond our near-term capabilities; traveling for eighty thousand years to another star is even more out of the question. In the future, expanding beyond our solar system is a worthy goal, but we need to start with closer destinations. As bases are established beyond Earth, there'll be enormous incentives to improve technologies and reduce costs.

Commercial initiatives to supply remote outposts will steadily make spaceflight cheaper and easier, so that someday it will seem no more daunting than air travel now.

So where are the best nearby destinations? At the top of the short list are the Moon and Mars. The Moon is much closer but also vastly inferior in long-term potential. It takes a few days to get to the Moon, or many months to Mars, but Mars is somewhat Earthlike, whereas the Moon is comparatively a dead rock. The main advantages of a Moon base would be ease of supply, and ease of rescue in case of emergency. There is some useful science to be performed—the Apollo missions explored only a tiny fraction of the surface, spending a mere twenty-four hours on surface operations across all six missions. It would also be possible to build highly sensitive radio telescopes on the lunar far side, where they'd be protected from humanity's incessant radio interference. One oft-cited reason to go to the Moon is to turn it into a giant gas station by synthesizing rocket propellant from its deposits of water ice. This idea has some merit, but the Moon isn't the miracle fuel depot it's sometimes made out to be. Most ice tends to be found in inaccessible locations such as the poles, and fueling and launch operations are difficult enough to perform on Earth, let alone at a lunar base lacking infrastructure.*

In considering destinations, it's important to ask ourselves what we're trying to achieve. The purpose of spaceflight is not just to reach space but to travel through it. Tourism aside, launching humans into orbit for the sole purpose of floating around makes little sense. We build ships to cross oceans and transport people, goods, and ideas. The same should be true of spaceships. To think of the Moon as a fuel depot is to admit it's a stepping-stone. But a stepping-stone to where?

---

* The Moon's *chief* advantage over Mars, it seems to me, is simply that it's more visible from Earth. Undeniably, the inspirational value of glancing up and saying, "People are living up there," would be powerful. (You can do the same for Mars, of course, but Mars tends to blend in with the stars, so, with less of a reminding presence, it's less likely to inspire on a nightly basis. Still, the strides that could be made on Mars to create a parallel Earthlike existence need only a good publicist to create equal, and, perhaps, surpassing, wonder.)

The answer is obvious: the only near-term destination that possesses the full spectrum of resources to support a permanent human presence is Mars. In the long run, we want to travel to other stars, but Mars is the logical next proving ground to see whether humans can establish an off-Earth existence. The fourth planet from the Sun, Mars is at the outer edge of the habitable zone, about 50 percent farther than Earth. With an equatorial temperature between 68 degrees Fahrenheit (20 degrees Celsius) and negative 94 degrees Fahrenheit (negative 70 degrees Celsius), Mars is cold, but not so cold that humans couldn't live in heated structures and explore using pressurized suits. At its equator, Mars receives about as much sunlight as southern Scandinavia or Alaska. However, the thin air means that far less heat is lost compared with equivalent temperatures on Earth. This effect is similar to the difference between water and air temperature: had the passengers of the *Titanic* been plunged into cold air instead of water, they would have survived for hours or days instead of minutes. You could comfortably walk around in a space suit on Mars, and someday—once the planet is terraformed—in just regular clothes. Compared with Mars, the Moon has no atmosphere, less than half the gravity (gravity is important for human health),* and fewer resources of any kind.

Mars possesses almost as much land surface as our planet.† Just like Earth, it experiences seasons, with a Martian year being about two Earth years long. Seasons on Mars are slightly more pronounced, since the planet has an axial tilt of 25 degrees compared with Earth's 23.5 degrees. A day on Mars is just slightly longer than a day on Earth, at twenty-four hours and thirty-nine minutes. Crucially, this means that plants could be grown in a Martian greenhouse using natural light. By comparison, on the Moon, plants would wither in fourteen-day lunar nights (the Moon has a twenty-eight-day day/night cycle). The Martian atmosphere is primarily carbon dioxide, which is important for growing plants, and can also be converted into oxygen and methane for rocket

---

* Lunar gravity and Martian gravity are around 17 percent and 38 percent of Earth's, respectively.

† Excluding oceans, Earth has only 11 percent more land surface than Mars.

fuel. Most noteworthy on Mars's list of assets is that it possesses abundant water. Although low temperatures and pressures mean that liquid water can't long persist on the surface, our robots have detected water seeping from underground. Someday Mars could be covered with lakes and oceans as it once was. With abundant water and the full spectrum of resources we find on Earth, Mars could support an entire human civilization.

There are at least three engineering options for getting to Mars. The first is to use multiple rocket launches to assemble one or more spacecraft in Earth orbit, which would then set off with everything needed to sustain operations all the way to Mars and back. There are numerous problems, however, with orbital assembly. Since all hardware would be designed and built on Earth over a number of years, if any launches fail, the timeline could be utterly ruined. Further concerns arise from having hardware sit too long in space, leading to potential reliability problems and the loss of propellant due to boiling and freezing. Assembling a large vehicle in Earth orbit is precisely how NASA had originally planned to get to the Moon, but the idea was replaced by a lunar orbit rendezvous in which a single Saturn V launched all the hardware directly to the Moon together. Considering that Apollo was already suffering schedule and cost overruns, a 1960s NASA fixated on massive orbital construction programs would never have made Kennedy's deadline—if it ever reached the Moon at all. While it's certainly *possible* to go to Mars by assembling giant spaceships in orbit, that method probably isn't the most efficient.

The second option to get to Mars is the minimalist approach. This relies on starting as small as possible with a few direct launches to the Red Planet, synthesizing methane-oxygen fuel on Mars to return.* With the minimalist approach, you'd want to produce as much as possible on Mars itself, including most bulk consumables, such as water, oxygen, and fuel. Martian "regolith" (soil) would be used to cover the living quarters as a shield from radiation, and regolith could also be made into bricks. Base modules would be pre-

---

* For more information on this "Mars Direct" plan, check out Robert Zubrin's *The Case for Mars*.

fabricated on Earth, with plans to expand by using local materials or by taking advantage of ready-made structures such as lava-tube caves. The minimalist approach relies on one or more launches to remotely pre-emplace modules and supplies before anyone travels to Mars (which is actually a good idea for *any* approach). Once a fully stocked base is established, settlers could travel to Mars with confidence, knowing that essentials awaited them on the Martian surface.

The third option is to build a big rocket that can loft a large interplanetary spacecraft in a single launch. To limit the spacecraft's mass to a practical level, it would probably be necessary to arrange for the spacecraft to refuel in Earth orbit before setting off to Mars. This would establish the capability to carry lots of people and cargo between the two planets, while ideally using some elements of the minimalist approach, such as methane-oxygen fuel produced on Mars. In September 2016, Elon Musk unveiled the basic architecture for this approach with the Interplanetary Transport System, consisting of interplanetary transports and refueling tankers carried by a very large rocket, all fully reusable (named in November 2018 as "Starship" for the spacecraft and "Super Heavy" for the rocket). Scheduled for subscale trials in 2019, this system could enable large-scale Martian settlement from the start. With these developments, it's now possible to imagine a sustained effort to settle Mars in the coming decades.

22 | BECOMING SPACEFARING

Our future has two possible paths. In one, we remain on Earth and eventually wipe ourselves out. In the other, we make the decision to become spacefaring. If our civilization were spread across many worlds, it would not only be insulated from disaster but also benefit from diverse and diverging interplanetary cultures exchanging technologies and ideas. The timeline and details are up to us, but expanding into space is inevitable if we're to survive. This is the culmination of a dream that dates back over two millennia to the Roman novelist Lucian of Samosata,* who wrote of extraterrestrial life and ships voyaging through space to other worlds. However, there's an important intermediate step, one often neglected in science fiction. If we make the decision to become spacefaring, we must expand into our own solar system first.

Our solar system is a big place. Most people learn the names of the planets, and possibly argue about whether Pluto qualifies for the honor of belonging to the club.† Yet, even for those who can name the major bodies circling the Sun, the solar system is a vague notion for most. Few people have any concept of how far away the planets are, what they're like, or how they interact with other nearby bod-

---

* Lucian was a Roman of Syrian origin who wrote in Greek.
† My view is that it doesn't matter. Pluto is what it is: a medium-sized chunk of ice just beyond Neptune. Lumping it into the category of "dwarf planet" is a good compromise because it expands the solar family to include other dwarf planets like Ceres, Haumea, and Eris.

ies. Even if they accept the vague idea that "eventually" we'll have to establish outposts on the more solid of these bodies as a hedge against the deterioration of conditions on Earth, that outpost building can often seem like an agenda for another decade, not ours.

To them, I would simply say: exploration is written into our genes, something that motivates us with common purpose, and there really is no downside to expanding into space. Throughout history, expansion has often come at the detriment of indigenous people or the environment, but expansion into space carries no such cost. In fact, there may be something of a windfall. Compared to some bodies in the solar system, the surface of Earth is a relatively resource-poor place. All the platinum and gold ever mined rained down on our planet long after it formed. Metals like iron and nickel are found in Earth's crust in trace amounts but are abundant in asteroids, including many that bombard our planet.* A single iron-nickel asteroid a mile across would yield more metal than we've mined in the history of civilization. Thousands of these are out there waiting for us.

Becoming spacefaring could generate enormous benefits for our home planet. Rare earth metals are essential ingredients in magnets, smartphones, computers, motors, medical imagers, nuclear reactors, wind turbines, solar panels, and batteries. Though not as rare as their name implies, they're found in extremely low concentrations. Extraction is expensive and environmentally damaging, since many tons of rock must be processed to extract small traces of metal. Rather than strip-mine vast tracts of Earth, we could gather these materials from asteroids, where they're found in enough abundance to satisfy demand for millions of years.

Dangerous and dirty manufacturing work could be transferred to orbiting space stations run by robots. Even food production could shift to orbit. Imagine, for example, hydroponic farms powered by solar energy that is many times more efficient than farming on our planet's surface, owing to the lack of regular nighttime

---

* Humans learned to smelt iron three thousand years ago, but iron weapons occasionally appeared before this, forged from meteoritic iron. A sword "falling magically from the sky" may be the genesis of sword legends like that of King Arthur's Excalibur.

intervals and an atmosphere that dissipates energy. With manufacturing, mining, and agriculture migrating off-world, large parts of Earth would return to a pristine state.

Our planet is a crowded place, but overpopulation is purely a matter of technology. The solar system has enough materials and energy to support between thousands and billions of times more humans than live on Earth today. The cost to extract most resources from space is currently prohibitive, but as we build far-flung settlements across the solar system, incentives and technological improvements will change this. Producing water, oxygen, rocket fuel, and other bulk consumables will be a crucial first step, but these ingredients are ubiquitous. All worlds in the solar system except Venus possess water, if only in solid form. In fact, many moons, asteroids, and comets contain more ice than rock, being mixtures of water, methane, carbon dioxide, and nitrogen compounds such as ammonia, all frozen. Some worlds, such as Jupiter's moon Europa, support vastly *more* water than Earth. And forget oil extracted from ancient underground fossils—Saturn's moon Titan boasts hundreds of times more hydrocarbons than Earth, filling entire lakes and oceans.

That we think of space travel as difficult is mostly due to the high energy cost associated with breaking free of Earth's gravity. If you started in space, space travel would be cheap and easy. Instead of being restricted to tiny capsules launched by giant rockets, commerce between asteroid bases could rely on interplanetary cruisers consuming only a tiny fraction of the fuel required to operate near Earth. In fact, leaving an asteroid would be so easy that you could travel to another by trampoline. Traveling between Mars and the Moon is more than *twice* as easy (in terms of fuel expenditure) as traveling to the Moon from Earth.* As for sending materials back home to Earth, that's easy if they can be harvested in orbit or tolerate a fiery descent through the atmosphere. For Earth, it's far easier to return than leave.

---

* This might seem strange since the Moon is so much closer than Mars (a three-day trip versus seven months), but since there is no drag in space, distance is inconsequential. All that matters is gravity.

So many opportunities open up when you don't have to contend with gravity. We've already landed robots on asteroids and comets—a major step toward using a spacecraft to nudge small asteroids into Earth orbit, where they could be conveniently harvested. We'd have to be cautious, though, given the damage that could ensue if a massive rock plunged into Earth's atmosphere. This highlights another reason to become spacefaring. Our planet sits in a shooting gallery of potentially lethal asteroids and comets. On average, we'll be hit by an asteroid large enough to demolish a city about once every century, and there's an outside chance of a larger impact that could destroy civilization altogether. Wouldn't it be nice to be able to *do* something about this? If the dinosaurs had had a space program, they'd probably still be here.

Several companies are getting into the asteroid mining business. Planetary Resources, founded in 2012 by a team that included XPRIZE's Peter Diamandis, film director James Cameron, and investors such as Google's Larry Page, is building space telescopes for prospecting and plans to build spacecraft for mining. The idea is to harvest metals to sell on Earth, and also extract water, oxygen, and hydrogen to support existing space operations. Another company, Deep Space Industries, was founded in 2013 with similar goals. With space mining, the initial cost is high, but the potential return on investment is huge. Even small Earth-crossing asteroids a few hundred feet wide—well within our technical ability to redirect—could contain metals worth billions. Profit margins could rival the sixteenth-century spice trade, when Portuguese navigators embarked on transoceanic voyages to reap hundred-fold returns. Some large asteroids are worth many times the gross domestic product of our entire planet.* If you seek abundance, look to the skies.

Once settlements are established across the solar system, they'll form an integrated space economy. From low-gravity worlds, materials could simply be hurled through space by electromagnetic rail guns. Such a catapult could be built on the Moon to fire

---

* The largest known iron-nickel asteroid, 16 Psyche, could supply Earth with metals for several million years at our current rate of consumption.

cargo all the way back to Earth. A similar system could be built by tunneling through Mars's Olympus Mons. Since this mountain is three times taller than Mount Everest, the tunnel would emerge above 98 percent of the planet's atmosphere, conveniently avoiding drag for materials fired into space. Another way to inexpensively transport materials into orbit would be a space elevator, a tethered cable extending from the surface up to a space station hovering above. Using motors, a platform could climb the cable to orbit (hence "elevator"). On Earth, a space elevator could theoretically be built from a point on the equator to a station in geostationary orbit; in the lower gravity of Mars or the Moon, space elevators could be built much more easily.

At a space settlement, there'd be huge incentives to operate sustainably. On the International Space Station, astronauts recycle urine and sweat, filter out the contaminants, and use the resulting water to rehydrate food, bathe, and even drink. Hydroponics and aeroponics (growing plants in water or mist) would be especially useful. The initial focus would be on growing garden crops such as lettuce, tomatoes, peas, beans, carrots, radishes, and strawberries as morale boosters to supplement bulk dried food. Potatoes, featured in Andy Weir's *The Martian*, are the most efficient food in terms of calories per area, so they're a great choice for space, along with sweet potatoes, which are among the healthiest foods. In fact, the so-called Irish diet of milk and potatoes provides almost all essential nutrients. The key missing ingredient seems to be molybdenum, consumed by the Irish in oatmeal form. For spaceflight, milk could be shipped as powder, with a few tins of oatmeal in the cargo hold. However, space food is usually chosen to suit preference as much as nutrition. For a homesick settler living at a remote outpost, freshly grown food would be an important reminder of the flavors of Earth.

You wouldn't bring animals on early flights, so meat consumption would be curtailed in favor of alternatives such as spirulina, a blue-green algae packed with protein.* Mushrooms could pro-

---

* In the early space program (a more meat-and-potatoes era), NASA studied the most efficient meat-producing mammals to bring. Mice topped the list.

vide essential B vitamins, with the added bonus of not requiring sunlight. And although it might offend North American sensibilities, another great dietary option would be insects. They're easy to raise in small spaces, grow quickly, consume waste products, and are among the most efficient animals known. As livestock, crickets are twelve times as efficient as cattle in terms of protein production versus food and water consumption, and they produce more nutritious meat with a far better ratio of omega-3 to omega-6 fatty acids.* A thousand species of insects are consumed by billions of people in two-thirds of the world. It's rather odd that Westerners eschew them while devouring similar delicacies such as lobster, crab, and shrimp.

A spacefarer would want to produce as much as possible locally, including hemp, bamboo, and other natural fibers that grow quickly and produce oxygen.† Plastics are another versatile option. Scientists have demonstrated the synthesis of bioplastics from plant waste, and ethylene can be produced on Mars and other worlds using atmospheric carbon dioxide and hydrogen electrolyzed from water. Ethylene is, in turn, the basis for most common plastics, including polyethylene, polypropylene, and polycarbonate, the latter of which can even be used to make transparent windows. Ceramics and glass are made from ubiquitous clays and silica. Perhaps most important, 3-D printing should be able to manufacture almost anything: instead of transporting a critical component, we'd just transmit the specifications at the speed of light. In the long run, everything we need can be produced using materials available elsewhere in the solar system.

Energy for a space settlement would come from highly efficient solar power (no night or atmosphere to contend with). Settlements farther out from the Sun could use nuclear power. In the long run, fusion reactors could potentially burn helium-3 to produce abundant power with no radioactive by-products. This isotope is

* For more on eating insects and potatoes in space, see my YouTube series *Cooking on Mars*.
† Hemp can make rope, cloth, paper, insulation, lightweight bricks, and biofuels, among other things. Bamboo makes an excellent wood substitute for furniture and structural support. Both plants have edible parts.

extremely rare on Earth, but it should be much more common in space, particularly embedded within the Moon's regolith (soil) and floating in the clouds of gas giants. Helium-3 reactions have been experimentally demonstrated, but it's unclear whether they could be scaled up or whether helium-3 could be economically extracted—particularly from the gas giants, where an elaborate network of buoyant stations and atmospheric scooping vehicles would be required.

Closer to the Sun, colonizing Mercury is a fascinating possibility. Near the poles, there should be deposits of ice (especially in the shadows of craters) and mountain peaks of eternal sunlight where solar panels could harvest copious amounts of solar energy (with a low axial tilt, the Moon also harbors such peaks of eternal sunlight). Venus, the hottest planet in the solar system, seems an unlikely homestead. Its stifling atmosphere is ninety times thicker than Earth's (more force per unit area than a polar bear dancing on a postage stamp), and the sulfuric acid rain is none too inviting. Yet, there's an altitude in the Venusian clouds where the atmospheric pressure equals that at sea level on Earth. Here, breathable nitrogen-oxygen air would float like a balloon in Venus's dense carbon dioxide atmosphere, so entire cities could be built floating in the clouds.* At cloud level, Venus is the paradise planet, where you could walk outside with no more than a short-sleeved shirt and respirator.

The outer solar system offers good prospects for settlement, but as you go farther out you get more ice and less rock. This means abundant water and rocket fuel, but heavier elements, such as metals, might have to be shipped in from asteroids. Refueling stations could be established at strategic locations throughout the solar system—for example, on the dwarf planet Ceres, a three-hundred-mile-diameter ice cube in the asteroid belt between Mars and Jupiter. A base on Ceres would be surrounded by asteroids, providing easy access to raw materials—although asteroids in the belt aren't as closely packed as we might imagine. More

* Like Cloud City from *The Empire Strikes Back* (although admittedly, I can't imagine what the purpose of such floating cities would be).

than a million asteroids are at least a mile in diameter, but since they're spread across thirteen trillion cubic miles of space, on average they're spaced farther apart than the distance from Earth to the Moon.

The four "Galilean" moons* of Jupiter have a good mix of ice and rock, and even trace atmospheres of oxygen and carbon dioxide (or in the case of Io, sulfur dioxide, which snows down on the moon's surface from volcanic plumes). Callisto and Ganymede are particularly attractive thanks to their distance from Jupiter's intense gravitational pull and radiation belt, making them easier and safer to travel to. The moons of the outer gas giants could also be settled, but they're so far from the Sun that nuclear power would probably be the only energy option. However, the majority of real estate in our solar system is farther still. The Kuiper Belt, beyond Neptune, may contain more than one hundred thousand icy worlds sixty-two miles (one hundred kilometers) in diameter or larger, the two most prominent being Pluto and similar-sized Eris. We're discovering more "trans-Neptunian objects" all the time, and an entire planet may be lurking out there in the dark. Even more mysterious are the trillions of comets of the Oort cloud, which occupies a region ten thousand times more distant from the Sun than Earth is. Our solar system is truly a vast place, endowed with enough resources and living space to support a spacefaring population of hundreds of billions.

The most intriguing place to settle is Saturn's moon Titan. The solar system's second-largest moon,† Titan is in many ways more Earthlike than Mars. It's the only other place with surface liquid. Covered with lakes of hydrocarbons, such as methane and ethane, Titan possesses hundreds of times the natural gas reserves of Earth. Titan also has the most Earthlike atmosphere in the solar system, composed of nitrogen and methane, just a bit thicker than Earth's. This means that humans could comfortably walk without

---

* Io, Europa, Callisto, and Ganymede: discovered by Galileo in 1610 orbiting Jupiter, thereby proving that Earth couldn't be the center of the universe.

† Ganymede is 2 percent larger, but Titan wins if you include atmospheric thickness. Titan is 50 percent larger than our Moon and slightly bigger than Mercury.

a space suit on Titan's surface—except that they'd freeze instantly in the negative 290 degrees Fahrenheit (negative 179 degrees Celsius) air. To make matters worse, heat loss would be more severe than for equivalent temperatures on Earth, so insulation presents the primary challenge, both for construction and for whatever extreme snowsuits Titan's inhabitants would wear. If the frigidity could be overcome, Titan possesses abundant resources, including vastly more water and carbon compounds than Earth. Amusingly, Titan's combination of low gravity and thick atmosphere means that humans could strap on wings and fly like birds simply by flapping their arms, and safely skydive without a parachute, reaching a terminal velocity of no more than fifteen miles per hour. Hovering vehicles would be trivial to operate, and buoyant cities could be built in the clouds.

Ultimately, one of the main reasons to go to space is to "terraform" one or more worlds to make them more habitable for humans. This might be possible for many places, including Venus and the Moon, but the most obvious candidate is Mars. Terraforming Mars would be returning it to an earlier state from billions of years ago, when it possessed oceans and a thicker atmosphere. When Earth and Mars were young, they were both warm and wet planets capable of supporting life. While our planet thrived, Mars became a cold, dry desert with temperatures too low and air too thin to support liquid water on its surface.* Why did Mars lose its atmosphere? On Earth, tectonic and volcanic activity constantly replenish the atmosphere, but Mars is far calmer. In addition, Mars's weaker gravity is less able to constrain atmospheric gases, a situation exacerbated by the planet's weak magnetic field, which allows high-energy solar protons to strip them away.

Terraforming Mars would mean reversing these effects by raising the planet's temperature and replenishing its atmosphere. We want to do on Mars precisely the opposite of what we want to do on Earth: create a runaway greenhouse effect. Mars has vast

---

* In fact, liquid water can currently exist on Mars within a narrow temperature range of around 32–50 degrees Fahrenheit (0–10 degrees Celsius). Below that it freezes. Above, it boils.

polar caps of dry ice. Raising its temperature would cause these to melt, releasing carbon dioxide into the atmosphere, which would trap heat from the Sun, raising the temperature further still. Mars has tremendous reserves of frozen water, in both its polar ice caps and its permafrost. With rising temperatures and pressures, liquid water should appear on the Martian surface, eventually forming lakes and oceans. Once the atmospheric pressure is increased by a factor of twenty (slightly lower than at the top of Mount Everest), people could comfortably walk around on Mars without space suits. Hardy microorganisms and plants could survive on a partially terraformed Mars, producing oxygen to eventually give Mars breathable air. (Indeed, some Earth bacteria could probably survive there today.)

How could this be done? The most permanent way to terraform Mars would be to stop the atmospheric leak, either by generating an artificial magnetic field* or by deploying a magnetic shield in space to protect the planet from solar wind. However, these steps may not be necessary. Mars lost its atmosphere gradually over several billion years—a very slow leak. With a current loss rate of under a pound of gas per second, if we could just replace the atmosphere faster than this, we could reverse the planet's decline into frozen desolation. A terraforming effort might deploy orbital mirrors reflecting sunlight onto the polar ice caps, absorptive materials sprinkled on the ice to increase thermal absorption, asteroids or comets redirected to impact the planet, nuclear detonations, or microbes genetically engineered to synthesize methane. Perhaps the simplest option is the one we seem to have mastered on Earth: release tons of highly potent greenhouse gases into the air. Chlorofluorocarbons (CFCs) are ten thousand times more potent than carbon dioxide, so with planet-wide factories producing them, it shouldn't take long to kick-start the terraforming process.

How long would terraforming Mars take? With transparent pressurized domes, we could terraform small sections of Mars

---

* This could potentially be done using a planet-wide network of superconducting rings.

rather quickly, giving humans a place to live and grow crops in Earthlike conditions. For planet-wide transformations, it's harder to tell and entirely depends on our dedication. As we start the process, we'll certainly invent better solutions, but this is how technology has always worked. Two centuries ago, we didn't know how to build heavier-than-air flying machines. It took smart people making incremental progress to work it out. Yet, unlike some challenges we face, terraforming Mars does not require major scientific breakthroughs: it's only a question of time, investment, and willpower. It may be an engineering project on a heretofore unprecedented scale, but we're pretty sure it's possible. Once we terraform Mars, other worlds will probably follow, so that one day our solar system will contain many worlds filled with diverse people, cultures, and civilizations, all descendants of a planet called Earth.

23 | GOING INTERSTELLAR

Traveling at highway speeds, it would take 50 million years to reach the nearest star, Proxima Centauri.* But suppose you could put your foot on the accelerator and boost your speed to over 38,000 miles per hour. How long would it take then? The most distant human-built object, Voyager 1, is traveling right now at that hard-to-imagine pace into the depths of space, but even Voyager 1 wouldn't get to Proxima for 75,000 years—*if* it were headed in the right direction (it's not). The fastest object ever built, NASA's Parker Solar Probe, is currently building up speed as it plunges toward the Sun and will eventually reach a breathtaking 430,000 miles per hour, hundreds of times faster than a speeding bullet. Yet even that mind-boggling pace represents only around 0.064 percent of the speed of light, and would get us to Proxima in no less than 6,500 years.

Travel to another star in a human lifetime quickly runs into a hard wall of physics. Kinetic energy is proportional to the square of velocity,† so reaching a speed a thousand times faster than the Parker Solar Probe—still a six-and-a-half-year trip to Proxima Centauri, not counting acceleration and deceleration—would require a million times more energy. Sending a spacecraft on a

---

* Nearest star other than the Sun, of course. A small red dwarf 4.2 light-years away (too dim to see with the naked eye), Proxima orbits the twin stars of Alpha Centauri every 550,000 years. In 2016, it was discovered that Proxima Centauri hosts at least one planet that could potentially support life.

† Ignoring propulsive and relativistic effects.

fifty-year voyage to another star would take more energy than is consumed in the United States during a year, somehow crammed into a container the size of a spacecraft. It's not simply a matter of adding more fuel. You have to carry the mass of whatever fuel you add, increasing the thrust required and resulting in a death spiral of diminishing returns. Conventional rockets are limited because they get their energy from chemical bonds, and you can only put so much fuel in a tank. If we want to get to another star in a reasonable time frame, we'll need alternative forms of propulsion.

What other types are available? Ion engines use electric fields to accelerate charged particles of propellant to extremely high velocities. Many interplanetary spacecraft have been equipped with them, and they're highly efficient because they can fire for a long time (often days or weeks) without consuming much fuel. However, ion engines do have some drawbacks. They have extremely low thrust (equivalent to breathing on a sheet of paper), consume a lot of electricity, and on an interstellar trip would still be limited by their propellant supply. Nuclear thermal engines are also possible. These heat propellants to high temperatures in a nuclear reactor and then fire the propellants through a rocket nozzle at high velocity. Nuclear thermal engines have never been used operationally, but several were tested from the 1950s up to the early 1970s, and were even planned for use on later variants of the Saturn V rockets that went to the Moon.*

A rocket's application of Newton's third law—for every action, there is an equal and opposite reaction—is not fundamentally different from the recoil of a gun. It's like pushing against water to swim forward, except in space there is nothing to push against, so we push against propellant ejected from a rocket instead. Chemical rockets push out vast quantities of propellant, generating a lot of thrust. Ion engines and nuclear thermal engines push out smaller quantities of propellant but at very high speeds, generating less thrust but maximizing efficiency over time. Nevertheless,

---

* Nuclear thermal engine development was canceled due to fears that a failure might disperse radioactive debris over a wide area. However, the danger was minimal since the engines used relatively small amounts of (nonexplosive) nuclear materials contained within strong cores to be launched over the ocean.

they're still limited ultimately by the size of their fuel tanks. This means that while they're great for cruising around the solar system, neither is truly suitable for travel to another star. Existing rockets just can't carry enough fuel.

What about using no engine at all? In a letter to Galileo written in 1610, Johannes Kepler observed that a comet's tail doesn't point away from its direction of motion, but away from the Sun. This implies that the Sun must be exerting some kind of "heavenly breeze" that could be captured to sail through the void of space.* Indeed, this is precisely what can be done. Solar pressure is measurable even on spacecraft without sails, to the extent that it must be accounted for in orbital and trajectory planning. However, the force is vanishingly small. The solar pressure captured by a solar sail a square mile wide would add up to less than a pound. Extremely thin films around the thickness of a human hair are used to make solar sails to minimize the mass. A large solar sail could propel a craft to Jupiter within a few years, but current sails are overdesigned because they have to survive launch to orbit and unfurling by mechanical means.

An ultrathin sail of lithium (the lightest solid element) could theoretically be built in orbit at a tenth the thickness of sails today—one five-thousandth the thickness of a sheet of paper. Using such a sail, a spacecraft could reach Pluto within a year or two. But for interstellar travel there is the problem that sunlight declines rapidly as a spacecraft moves farther from the Sun, and in pushing beyond our solar system a spacecraft quickly gets too far away. Suppose, though, that we could create our own "wind"? By aiming a laser at a sail out in deep space, we could push it along even once the sunlight had faded. The craft would have to be small, and it could take a lot of energy to focus a laser far enough into space. At best, we might get to Proxima Centauri in fifty years with a tiny probe—but that would require a sail sixty-two miles across pushed by a laser consuming twenty-six thousand gigawatts: around double Earth's entire power generation.

There's only one existing technology that might get us to

---

* Kepler to Galileo: "Given ships or sails adapted to the breezes of heaven, there will be those who will not shrink from even that vast expanse."

another star in a reasonable time frame, but it sounds a bit wacky. Back in the 1950s, a group of scientists studied the possibility of using nuclear bombs to propel a spacecraft. Called Project Orion, the idea was simple: you push nuclear bombs out the back and ride the detonation shockwaves on a specially designed pusher plate. The advantage is that your acceleration comes not from chemical bonds but directly from nuclear reactions, thus liberating millions of times more energy. There's no upper size limit for ships powered by nuclear blasts, since a larger ship could just carry more bombs and better survive the shock waves. The scientists who worked on Project Orion envisioned interplanetary and eventually interstellar ships the size of cities, but the project was killed in 1963 by a treaty banning nuclear weapons in space—and also, probably, by the fact that even Apollo's price tag was tough to swallow, let alone the anticipated cost of city-sized nuclear spaceships. Yet, the basic principle was sound, and such a ship might be able to reach 5 percent of the speed of light, getting to another star in around a century: still a long time to wait, but at least a human lifetime, more or less. There is, of course, the problem of launching large numbers of miniaturized high-yield nuclear weapons into space, an activity that seems self-evidently hazardous. While this method of propulsion is *technically* feasible, it's hard to imagine the world's politicians signing off on the idea any time soon.

Assuming we're unwilling to ride nuclear bombs to the stars, what else is on the horizon? Fusion power has the potential to generate ten million times more energy than chemical reactions. Theoretically, fusion power could solve all of Earth's energy problems, providing a sustainable, clean energy source. The only problem is that we haven't quite gotten fusion power to work yet—at least, not to the point where we can produce more energy than what must be put in to start the reaction. It's one of those technologies that's always twenty years away—but if we do get it to work, we could use fusion to power rockets. Depending on the type of reactor used, a fusion rocket could either produce energy to drive a very efficient ion engine or simply direct the fusion reaction exhaust products out the rocket's rear. Someday, fusion rockets could get us to another star, but the trip would still probably take decades and the rockets would have to carry a thousand tons of hydrogen fuel: the mass of a small navy warship, or the yearly launch capacity of all nations on Earth.

Or, instead of carrying hydrogen to fuel the fusion reactor, why not collect it along the way? Although space is extremely sparse, it's not completely empty. Interstellar space contains around one atom per cubic centimeter. This is less than a quintillionth the density of air (1 followed by eighteen zeroes), but, with an enormous scoop, hydrogen atoms could be collected and concentrated into a fusion reaction. Proposed in 1960 by physicist Robert Bussard, this mega-scoop, "Bussard ramjet"* would use a magnetic field several kilometers wide to gather stray hydrogen atoms into its enormous maw. Although this technology could theoretically work, it's unclear if deep space has enough atoms to sustain fusion reactions, or if the ramjet could overcome the solar wind of the star it was traveling toward. So fusion rockets? Probably, yes. But interstellar fusion ramjets are a definite maybe.

A thousand times more energetic than fusion power is antimatter, whose reaction with normal matter results in the maximum possible conversion of mass to energy (Einstein's $E = mc^2$). It sounds strange, but every particle has an opposite "antimatter" equivalent with the same mass but opposite quantum numbers. These atomic doppelgangers are created sporadically in nature but don't last long because matter and antimatter instantly annihilate each other when they meet, releasing an intense burst of energy. As with fusion power, we could use antimatter to power an electric engine, but to achieve maximum efficiency, we'd have to find a way to direct the reaction products out the back of a rocket to produce thrust.† Since antimatter exhaust would travel at relativ-

---

* A ramjet is an engine (usually on an aircraft) that uses forward motion to compress incoming air, meaning it can only be used at high speeds and can't start at a standstill.

† *Star Trek* ships use antimatter, but to power a warp drive, not generate thrust directly. Well, most *Star Trek* ships do. Romulan ships are powered by an artificially generated singularity, i.e., black hole. Both types of power sources have an inconvenient habit of exploding in a rather intense way when they lose containment (or, rather, imploding in the Romulan case).

istic speeds, rockets could accelerate to a significant fraction of the speed of light. This would allow travel to another star in a matter of years—if we could solve the production, containment, and lethal gamma ray problems. Currently we can't produce more than a few atoms of antimatter, and they quickly react to annihilation with, well, everything, because everything else is normal matter. Antimatter is the most expensive product in existence. NASA estimates that it would cost $60 trillion to produce a single gram of the stuff, and we'd need a lot more than that.

Is there any way to get to another star faster? There isn't within our understanding of the laws of physics, but that doesn't mean there never will be. One way to get around the awkward problem that nothing can move faster than light would be to keep a spaceship stationary but "warp" space around it. This is the solution adopted by *Star Trek*, in a vague notional sort of way, but it is consistent with at least one property of the universe. The universe is "only" 13.8 billion years old, but it has a visible diameter of around 93 billion light-years. This seems to imply that parts of the universe somehow moved away from each other faster than the speed of light—but what it actually means is that space itself is expanding, not from a central location, but in many (possibly infinite) locations. (Picture a ball of dough with raisins as "galaxies." When you bake the dough, it stretches out, moving the "galaxies" apart.) Since space can expand and compress, instead of traveling faster than light, could we ride a bubble by compressing space in front of a ship and expanding it behind?

In 1994, Mexican physicist Miguel Alcubierre proposed a way to create a warp bubble by producing an energy-density field lower than that of a vacuum. This could theoretically work, except it would rely on large quantities of either negative energy or negative matter, which may not even exist.* There are other complications, including the inability to steer or stop the ship because signals could never penetrate the bubble, and the fact that Hawk-

---

* Negative matter is confusingly not the same as antimatter. Antimatter has been created in a lab, but negative matter has never been observed. Negative energy has been generated in extremely minute quantities.

ing radiation would obliterate everything inside.* Even Alcubierre now thinks his idea is impossible, although we can't entirely rule out the chance that these barriers will someday be overcome. But even if we can someday create a warp bubble in a laboratory, this would be a far cry from turning it into a practical transportation system. Thus, for now "warp drives" must remain in the "maybe someday, maybe never" category.

How long would it take to get to interstellar velocities? If we could sustain a steady acceleration equivalent to Earth's gravity (1G), it would take less than a year (354 days) to accelerate to the speed of light.† This modest level of acceleration is less than a fifth of that experienced during a rocket launch, and if it could be sustained for long periods of time, interstellar travel becomes possible. Our galaxy is vast, consisting of four hundred billion stars spread across a hundred thousand light-years, but even if we never approach the speed of light, humans could one day expand across it. At a mere 5 percent of the speed of light, we could settle the entire galaxy in two million years: a long time, but not much longer than our species has existed. Even at the fastest speeds we can reach today, we could travel across the galaxy in a few hundred million years: less than a tenth of the age of our planet, and not much longer than mammals have inhabited Earth. Still, a long time from the perspective of a human life.

Even at small fractions of the speed of light, there are some problems with interstellar travel. As ships accelerate to extreme velocities (with correspondingly high energy levels), individual atoms pose a deadly threat. Indeed, even at orbital velocities of several miles per second, small pieces of debris can rip through metal plates. (The International Space Station's windows are scarred with streaks where tiny particles have cut the glass, and astronauts report hearing "pings" when tiny projectiles tear through the station's solar pan-

---

* Hawking radiation is the effect that allows black holes to emit radiation even though light can't escape.

† If the floor was oriented toward the back of the rocket, this acceleration would feel exactly like standing on Earth (the same effect as being pulled down in a rising elevator). Of course, accelerating all the way to the speed of light would be impossible, but it goes to show that continuous acceleration adds up fast.

els.) Thus, for relativistic travel, we'll need to devise a way to deflect sparse interstellar atoms.* This might be possible using some kind of laser to ionize atoms so they can be deflected with a magnetic field or channeled as with the Bussard ramjet. Another problem with interstellar travel is that communicating with a spacefarer's home planet is tough since radio signals decay with the square of distance. Radio antennas two hundred feet wide, part of NASA's Deep Space Network, are required to communicate with the New Horizons spacecraft out past Pluto. Proxima Centauri is thousands of times farther, meaning radio signals would decay by factors of millions. One way to get around the signal loss would be to communicate with extremely high-powered lasers, or even send back physical memory packets on tiny spacecraft—but it would still take years to transmit a message, let alone receive a round-trip reply.

How could we keep a crew alive for the many—possibly hundreds—of years it would take to get to another star? One option is to send a "generation ship," where only the great-grandchildren of the people who set out would reach the destination. This could entail building a giant starship, or hitching a ride on an interstellar comet, whose materials could be harvested to build, sustain, and propel a space colony traveling to the stars. Or we could make pit stops at the many "rogue planet" islands floating in the space between stars. There may be many times more rogue planets than regular ones (several billion in our galaxy), but they're tough to find. So far, around twenty are confirmed or strongly suspected. Dark and distant from any sun, some may nevertheless have moons that are heated by tidal forces from the host planet, like Jupiter's moons, and might be able to support life. Like the vagabond comets plying the space lanes between stars, rogue planets could act as refueling stations or natural spaceships, furnishing resources for the journey. Multigenerational voyages aren't entirely unprecedented. The humans who left Africa to eventually settle in North America took perhaps fifty thousand years to complete the journey, or several thousand generations.

* *Star Trek* solves this problem with the oft-referenced but never explained "navigational deflector."

Or perhaps we could preserve the crew? Creatures such as tardigrades,* some insects, and some types of turtles and frogs can revive after being frozen. It's unclear if we could do this with humans, but this hasn't stopped hundreds of people from having their bodies frozen at the moment of death in case they can ever be safely thawed (presumably after a cure is discovered for whatever killed them). Easier than freezing would be hibernation. This would reduce body temperature, slowing metabolism to a comalike state where the crew could be monitored and fed intravenously. Many mammals and birds survive like this throughout the winter, and hypothermic therapy is already being used by hospitals to extend the life of patients in cardiac arrest for hours or days. An alternative to sending a preserved crew would be to send frozen embryos, to be raised on their new home planet by robot guardians. Combine preservation with fusion or antimatter rockets, and it's possible to imagine a voyage to another star being initiated by the end of the century.

If humans are to survive in the long run, we'll have to travel to another star. But without a major breakthrough in our understanding of the laws of physics, the challenges of interstellar travel mean that a voyage will take an awfully long time. Instead of hopping from star to star like in *Star Trek* or *Star Wars*, our future may include gargantuan interstellar colony ships setting out on epic one-way voyages across the depths of space. In the meantime, we're already scouting for landing sites, as our telescopes scan the skies for planets around other stars. Soon, we'll start sending robotic explorers, and this could enable virtual participation. Imagine simulated vacations to a distant point in the galaxy, perfectly reconstructed based on data transmitted by robotic pathfinders trillions of miles away. Yet, as enticing as this prospect is, I suspect it won't satisfy us forever. We come from a long line of explorers. Someday a band of bold pioneers will leave the warm embrace of our Sun, if only because restlessness is part of being human.

---

* Tardigrades resemble microscopic caterpillars. Found everywhere from mountaintops to the deep ocean, they tolerate extreme temperatures and pressures, lethal radiation, and even the vacuum of space.

24 | LIFE ON OTHER WORLDS

Observations by the Kepler Space Telescope suggest that there are billions of Earthlike planets in our galaxy—at least one per star, on average. Across billions of galaxies in the universe, this would imply more Earthlike planets exist than all the grains of sand on all the beaches of Earth, around 1,000,000,000,000,000,000,000,000: a septillion or more. If we include the possibility of life in exotic environments such as the subterranean oceans of icy moons, the number of worlds capable of supporting life should be ten times higher. We can therefore be certain that there is plenty of habitable real estate out there. As we've seen, even if we could never travel faster than 5 percent of the speed of light, we could expand across the galaxy in less than two million years. The galaxy has been around for 6,500 times that long—around 13.5 billion years—so if a spacefaring extraterrestrial civilization had arisen at any point during that time, they should have overrun the galaxy, including Earth. So why are we here, and why aren't they instead?

This question is known as Fermi's Paradox, after the Italian physicist Enrico Fermi, and it's actually rather puzzling. Even if extraterrestrials had some bizarre aversion to space travel, shouldn't we detect signs of their presence? Where are the radio signals from space? Several solutions have been proposed to resolve this conundrum, but none is particularly satisfying. Perhaps life is unlikely to start in the first place, or give rise to intelligence? Perhaps beings on other planets have no interest in our world? Perhaps they're so far away that the vast distances have so far prevented contact? Perhaps truly advanced beings stay home, finding other ways to entertain themselves? Perhaps they've transcended their biological form, uploading their consciousness into computers? Perhaps we're simply too dull or stupid for them to bother with? Or perhaps they just haven't noticed us yet? The universe is, after all, a pretty big place, and even telescopes have peered into no more than a tiny corner of our own galaxy. Concluding that there is no life because we haven't found it yet is, as Neil deGrasse Tyson observed, like examining a cup of seawater and concluding that there are no whales in the ocean.

In 1961, the astronomer Frank Drake suggested a way to estimate the number of detectable extraterrestrial civilizations based on the number of habitable planets, the chance of intelligence and technology arising, and the life span of civilizations.* The problem is, until recently, we had nothing beyond wild speculation to feed into the Drake Equation, and the result varies widely based on assumptions. In recent years, with the search for planets outside our solar system, we're finally getting a fix on some of the equation's variables. Around two stars are created every year in our galaxy, and most of these have planets. Tens of billions of these should be habitable. It seems likely that intelligence eventually gives rise to civilization. The tricky terms are the probability of life and then intelligence arising, and the life span of civilizations. Depending on what values you assume for these, the Drake Equation returns solutions that vary from just one (us) to millions of civilizations in our galaxy. Considering the ramifications of Fermi's Paradox, the real answer is probably on the lower side. But that doesn't mean life in the universe is necessarily rare.

Earth is quite a lovely rock, the most habitable world we know, but this doesn't mean that it's part of an exclusive club. It's not that we magically dropped out of the sky onto a planet that hap-

---

* Number of intelligent civilizations = Rate of star formation × Fraction of stars with planets × Number of habitable planets per star × Fraction of these planets that develop life × Probability of intelligence arising × Probability of developing communications technology × Life span of civilization.

pened to be perfect for us. We've been shaped by billions of years of evolution on this planet. Earth seems so amazingly habitable to us precisely because we evolved to live on it. If we evolved on a planet bathed in ultraviolet radiation, we might find Earth's ozone layer intolerable because we'd never be able to soak up an adequate dose. Not only have Earth creatures been shaped by the planet's environment, but the reverse is also true. The very rocks beneath our feet are composed of layers of long-deceased marine organisms. Our civilization burns the remains of ancient life in power plants, cars, and airplanes. When life first appeared, oxygen was rare. Then, around 2.4 billion years ago, some types of microbes began to produce oxygen as a toxic by-product. Eventually, when the planet's atmosphere became saturated with the stuff, new forms of life adapted to breathe it. Organisms change their environment as they evolve to suit it.

Earth was not always so hospitable. When life first arose, our planet was a searing, radioactive, volcanic hellscape pummeled by incessant impacts from space. Geologists call this the Hadean era, after the Greek underworld. Our planet has experienced numerous phases, possibly including a "snowball" period lasting hundreds of millions of years where the oceans froze over and, from space, Earth would have appeared as lifeless as Jupiter's moons. Life persisted. Even now, life flourishes in an extremely wide range of conditions. Bacteria live essentially everywhere from the upper atmosphere to the depths of the planet's crust. They tolerate intense radiation, temperature and pressure extremes, abundance or lack of light, and utter deprivation of water and nutrients. They bathe in acidic pools of scalding water, dining on chemicals oozing from volcanic vents. They thrive deep underground in oil wells and trapped miles under the Antarctic ice sheet. Some bacteria live out their lives encased within rocks, subsisting on nothing but minerals. In fact, there's more life within Earth than on top of it, with up to a hundred trillion tons of bacteria below the surface. Pile them up, and they'd cover our entire planet to a depth of over five feet.

Life does have limits. We wouldn't expect to find life on the surface of a star because the intense heat would vaporize complex molecules.* So far as we know, there must be a temperature range capable of supporting stable organic compounds, and (we think) some sort of liquid to act as a solvent. Water is ideal but may not be the only substance capable of serving this purpose. The surface of Saturn's moon Titan is covered in lakes of hydrocarbons at temperatures colder than liquid nitrogen, but we can't rule out the possibility that microorganisms or even larger creatures could occupy such frigid environments. If they do, they must have different chemistry (life, but not as we know it). However, since we know of only one world that definitively supports life, it seems safe, if somewhat chauvinist, to assume that conditions for life are best represented by the diverse array of environments found on Earth.

Life began on our planet soon after it became habitable, but for more than 80 percent of the time life has existed, it has been strictly single celled. This suggests that life in the universe may be relatively common, but complex life may be much rarer. There may be many worlds out there swarming with bacteria, or oceanic worlds teeming with algae, but perhaps not many worlds with fish, lizards, trees, goats, or lions. Yet, the observation that life began early in Earth's history may not mean as much as it appears. The origin of life could have been a random fluke. As an experiment, a one-out-of-one result is virtually meaningless, because we need to be here to make the observation. If we were the only life in the universe, we'd have no way of knowing, because the billions upon billions of lifeless worlds out there would have no one to make a competing observation.† But, on the other hand, if we were to find life anywhere else, even in our own solar system, and it didn't share a common origin with Earth life, this would imply that life is abundant throughout the entire universe.

We have plenty of places to search for life. At least ten solar

---

* Unless the star had died long ago and cooled down since, but no stars have had enough time to do this. It would take trillions of years for a star to get cold enough to touch, and even then its intense gravity would still kill you.

† Called the "anthropic principle," this can be extended to observe that we can't know if the laws of physics have been "fine-tuned" to allow for life, because only universes that could develop life would be observable by that life.

system worlds are thought to possess liquid water below their surface.* Saturn's moon Enceladus erupts geysers into space that freeze into showers of misty ice crystals. NASA's Cassini spacecraft sampled salt in the plumes, hinting at warm subsurface reservoirs not unlike Earth's oceans. If life developed on Enceladus, it's likely that microorganisms would be ejected with the water, waiting to be scooped up by a passing spacecraft. Jupiter's moon Europa offers even better prospects for life, with an ocean beneath its ice sheet containing more water than all the oceans on Earth. Europa's ocean should be in direct contact with the bedrock below, and since the moon is flexed throughout its orbit by Jupiter's gravitational field, intense tectonic activity should feed volcanic vents on the seafloor. Spewing out warm nutrients, volcanic vents in Earth's oceans are productive environments, analogous to the ancient ecosystems that spawned life on our planet. Could microbes or even multicellular creatures be swimming in the darkness of Europa's oceans?

Considering that subterranean reservoirs exist on Mars and that the planet was once warm and wet, it seems likely that Mars once supported life, too. Since it took millions of years for Mars to become a cold, dry desert, is it not likely that pockets of life would survive, either underground or by adapting to the new conditions? The jury remains out for the case of life on Mars. We've only looked for life twice, with the pair of Viking landers in 1976, each conducting four astrobiology experiments. Three out of four experiments came back negative, but the last yielded a positive result.† Organics (essential ingredients of life) were not detected, but this could have been due to the presence of perchlorates, which have been detected in high concentrations by recent missions. When

---

* Mars, Europa, Enceladus, Ganymede, Callisto, Ceres, Titan, Mimas, Triton, and Pluto.

† The four experiments were: 1) a mass spectrometer looked for organics; 2) a chromatograph looked for the production of oxygen, carbon dioxide, and other gasses; 3) a photosynthesis experiment looked for carbon fixation; 4) nutrients were added with marked carbon-14 to look for metabolic activity. This last experiment was positive for the first injection of nutrients, but further injections did not register any response (suggesting that either the initial positive was from an unknown chemical process, or that organisms were present but they had perished).

heated, perchlorates react with organics to produce chloromethane and dichloromethane, which *were* detected by Viking—suggesting that organics may have been destroyed before detection. Indeed, when researchers added perchlorates to Marslike soil from the Atacama Desert and repeated the Viking experiments, they got the same results. So maybe there's life on Mars after all?

Even if the Viking results really were negative, extrapolating this to conclude that there is a total lack of life on Mars would be like sampling a patch of soil in Texas and another in Siberia and concluding that no life exists on Earth. In reality, the search for life on Mars has hardly begun, and with each mission we learn more about potential subsurface sanctuary ecosystems. Not that sanctuaries are necessarily required—just look at the extreme conditions that life tolerates on *our* planet. Certain strains of bacteria from Earth could probably survive on Mars's surface today. This is especially fascinating because fragments of Mars are occasionally blasted into space by impacts, traveling through space to hit our planet as meteorites. At least 132 meteorites collected on Earth originated on Mars. Could a few of those Mars rocks have brought life along with them in Earth's distant past?

This idea that life hitched a ride through space ("panspermia") simplifies life's origin because it would only have to begin once and then hop around the cosmos on interstellar comets. Did we originate eons ago, on another world circling a different sun? Maybe, but we don't need to rely on this possibility to explain the origin of life. Although we wouldn't expect organisms to suddenly spring into existence out of nothingness, there's no reason to suppose that life is especially rare. The building blocks of life, organic molecules and amino acids, are ubiquitous, frequently raining down on our planet attached to meteorites. Nor are amino acids difficult to synthesize—in fact, they'll spontaneously assemble in an electric charge, as demonstrated in a 1952 experiment by Stanley Miller and Harold Urey. Mixing water, methane, ammonia, and hydrogen in a flask representing Earth's primordial soup, Miller and Urey subjected the concoction to electric sparks representing lightning. Within a week, all the amino acids found in life had emerged.

What is tough is assembling proteins, which are often composed of hundreds or thousands of amino acids. It's not just the number that makes proteins complex but the fact that they have to be ordered in the right way and folded into tangled three-dimensional shapes to do their jobs. It's a big step to move from amino acids to proteins, one that the astronomer Fred Hoyle* compared to a tornado sweeping through a junkyard and spontaneously assembling a Boeing 747. It does seem unlikely that proteins would suddenly spring into being, and they probably didn't. But there's no need to start on such a grand scale. Basic life merely needs to store information, drive chemical reactions to extract energy, and reproduce. In our cells, DNA stores information, while proteins direct chemical reactions. But our cells also use a simple form of nucleic acid called RNA to act as a short-term messenger between DNA and proteins. Unlike DNA, a wound-up double helix, RNA is composed of simple chains that could spontaneously assemble. Conveniently, RNA can perform the functions of both DNA and proteins. In fact, RNA does precisely this in certain primitive organisms, such as viruses. So we can imagine life starting as simple self-replicating RNA chains, which eventually evolved to produce proteins and DNA for specialized tasks.

Once organic molecules reproduce themselves with possible variations at each generation (mutation), you have evolution by natural selection. Evolution is the only means by which complexity can emerge, by gradually accumulating tiny changes over millions or billions of years. Because mutations are random, it's sometimes supposed that evolution is random, but in fact it's precisely the opposite. Evolution is a directed means of finding solutions by trial and error. With each generation, variations that improve reproductive success tend to get passed on, while those that don't are discarded (by death). Imagine a population of animals who chase down prey. Random mutations may produce a variety of body plans, but only the strongest and fastest survive. Eventually, the entire population is equipped with big muscles and

* Fred Hoyle came up with the term "Big Bang," although it was meant as a derisive rejection of the idea.

streamlined bodies, exactly as if they'd been designed for running. Evolution will mold extraterrestrial life, as surely as physics molds distant planets, stars, and galaxies.

What would alien life be like? At first, it seems as if it would be *very* different. Imagine life on a planet where one side is always facing the sun.* The sun side would be too hot, and the night side would be too cold, but there might be a ring of life around the constant horizon. Or imagine gas giants like Jupiter with colonies of floating bacteria feeding ecosystems of flying creatures soaring and diving through the clouds. Given unique environmental pressures and the vagaries of circumstance, natural selection doubtless produces a kaleidoscope of fascinating creatures. Yet, biological functions might not be so different. The laws of physics are the same everywhere in the universe. While it's not impossible that life would use different molecules—silicon, say, instead of carbon—there are good chemical reasons why Earth life is the way it is. Like carbon, silicon can bind four other atoms and form chains. However, silicon chains are more fragile, and while we exhale carbon dioxide to purge waste, it's less convenient to breathe out solid blocks of silica.

Have you ever noticed that vertebrates (fish, reptiles, amphibians, dinosaurs, birds, mammals) are rather similar? They come in all shapes, sizes, and colors, but they all have a head, body, four appendages (arms, legs, wings, or fins), digestive tracts running through their bodies, and symmetric left and right sides. This is because all vertebrates share a common ancestor. Before around 540 million years ago, the animal kingdom consisted of coral, burrowing worms, jellyfish, and not much else. Then, during the "Cambrian explosion," there suddenly (in geological terms) appeared a dazzling array of animals with eyes, legs, fins, and feathery gills. For the first time, hunters chased down prey, crushing it between toothed jaws. All complex animals on Earth trace their ancestors back to this time.

In 1909, the best-preserved collection of fossils from this period

* "Tidally locked," where a world's rotation rate is the same as its orbital period, like our Moon.

was discovered by Charles Walcott* in the Canadian Rockies. Called the Burgess Shale, not only does this fossil bed contain an astonishing variety of creatures, but they're exquisitely preserved, with imprints of eyes, tissue, and other soft body parts. When he first discovered the fossils, Walcott tried to sort them into existing classifications by comparing them to living creatures, but many of the stranger specimens never quite seemed to fit. For a long while, they were regarded as mere curiosities. This metaphorical shrug existed for half a century, until scientists in the 1960s started to take a comprehensive look at the sheer diversity of life represented by the fossils. Some of the creatures were well-known, such as trilobites—which roamed the oceans for 270 million years up to the dinosaur age—but others were downright peculiar, with bizarre features unknown to science.†

Some of these creatures had five eyes. Some had arms emerging from their heads. Some projected flowerlike stalks to filter-feed in Earth's ancient seas. There were giant predatory shrimp, clawed sea scorpions, and spiky worms several feet long. Many seem to have been evolutionary whims with no modern-day descendants. But buried among these impressive creatures was the small, humble, eel-like *Pikaia*, sporting the first primitive spine. Nothing suggested that *Pikaia* was anything special: it wasn't the largest, strongest, or most numerous creature. And yet, its lineage would give rise to most of the animals we're familiar with, including us. Was *Pikaia*'s triumph a sheer accident, or was there some reason for it? What made its body plan superior? As the biologist Stephen Jay Gould asked, if we were to turn back the clock of time and play it over again, would we get the same result?

The answers to these questions could tell us a lot about life on

---

* In 1916, Walcott awarded Robert Goddard his first rocketry grant, so he also had a hand in getting us into space.

† Confronted with the sheer diversity of specimens, an investigator named Simon Conway Morris famously exclaimed, "Oh fuck, not another new phylum!" A phylum is the classification level right below kingdom. There are thirty-four phyla of animals, but most you're probably familiar with (mammals, birds, fish, reptiles) are Chordata. Insects, spiders, and crabs are Arthropoda. Squids and snails are Mollusca. The remaining thirty-one phyla are mostly biologically diverse but superficially similar worms.

other worlds. Given similar conditions, will evolution find similar solutions, or will it sculpt unrecognizably alien forms? We don't know for sure, but we can say that a few key breakthroughs had major impacts. More than perhaps any other organ, eyes have shaped the evolution of creatures on Earth. Sensory organs such as eyes allow animals to hunt, evade predators, and actively search for food. Without them all you can do is drift about aimlessly, hoping that food somehow lands in your mouth. But eyes come in various forms. Many microorganisms have receptors with the ability to detect the direction and intensity of light. Even complex eyes with the ability to focus are reckoned to have evolved independently on our planet more than fifty times. Humans are visual creatures, but our eyes are far from the best. Nocturnal animals have far better night vision, birds of prey have far better distance vision, and some animals can see a much wider range of colors.*

Animals evolved eyes on Earth because they're useful here. Our Sun is a giant fusion reactor leaking vast amounts of radiation into space in the form of photons. Many of these photons hit our planet and bounce off of things. Evolution stumbled on the fact that the energy from photons stimulates certain proteins, converting light into a biochemical signal. Photoreceptors can be used to detect colors because they differentially absorb or reflect different wavelengths of light.† Since most habitable worlds orbit stars, it stands to reason that most complex life should develop eyes. But not all stars are the same. If we lived around a dim star like a red dwarf, we'd probably see deep into the infrared spectrum. The ranges of light on Earth are visible to us because we're equipped with sensory organs that evolved to detect them. Even on Earth, there are wide varieties of eyes. Tuataras‡ from New Zealand have a third eye on top of their heads, and some animals keep eyes in

---

* With a dozen types of color receptors, mantis shrimp, for example, can see colors well into the UV range. (We have three types of receptors, dogs have two.)

† In a sense, objects are the "opposite" of the color you think, because you see reflected light. For example, a green object reflects green light, meaning that it preferentially absorbs colors other than green.

‡ Tuataras look like lizards, but they're actually from a separate lineage predating most dinosaurs.

stranger places. Sea creatures called chitons have light-sensitive "eyes" adorning the entire surface of their shells. Some types of tube worms have eyes on their feeding tentacles. Box jellyfish have twenty-four eyes of four different types, giving them three-hundred-sixty-degree vision.

Yet even though we find an incredible diversity of life on Earth and should expect to encounter the same on other worlds, there may be good reasons why *intelligent* beings are not so very different. For one thing, intelligent organisms probably need to be a certain size to be able to support a large brain. Insects are one of the most successful animal classes on Earth, but we're not likely to find intelligent insects because there are fundamental upper limits to their size. Insects breathe through tiny holes in their skin, and since larger animals have lower surface area–to–volume ratios (surface area is proportional to the square of size, volume is proportional to the cube), a giant insect would suffocate (doubling an insect's size would halve its relative oxygen intake).* We humans get around this problem by using lungs equipped with millions of alveoli that collectively boast the surface area of a tennis court. Large insects might also take a long time to shed their exoskeleton during molting, leaving them vulnerable to predation (not to mention the resources wasted with each discarded skeleton). This is already a problem for large lobsters on Earth, and intelligent insects might have to be much larger.

What about body plan? Evolution often converges on the same solutions, even from different directions. Fish, mammals, and reptiles are very different, yet sharks, dolphins, and prehistoric ichthyosaurs all have similar body plans that evolved independently. On Earth, we have two major types of symmetry: bilateral, where the left and right sides of the body are mirror images (like us), and radial, where a pattern forms around a center, like a starfish. These symmetries are efficient for balance and because genetic code can be reused, and it's a good bet that life on other worlds would use them, too. If evolution were rerun, we could have five eyes and three ears. But would we?

* The largest insect today is the New Zealand giant weta, a cricketlike monster up to eight inches long. But during dinosaur times, Earth had more oxygen and was covered with giant insects like dragonflies with wingspans several feet wide.

We have two of many things, partly due to bilateral symmetry, but also to provide redundancy (if you lose one eye, you can still see). We only have one heart and stomach because these are costly to create, and a pair might get muddled up trying to work in tandem.

Organisms need energy, so they'll have some sort of stomach and digestive tract, and a circulatory system to move nutrients and waste around the body. This could take different forms—starfish, for example, eject their stomachs out of their bodies to feed—but there does have to be a way of getting food into the body. Organisms also have to send commands throughout their body, so we can add a central nervous system to the must-have list. It makes sense for the most important sensory organs, such as eyes and ears, to be near the brain, pointed in the direction they would interact with the world. The sensors should be mounted high up in order to detect opportunities and danger from afar, and the compact brain-sensor package might as well be enclosed in a protective casing. So now we have digestion, circulation, nerves, eyes, ears, and brains protected by skulls mounted on necks.

Intelligent beings would communicate, to coordinate actions and exchange knowledge. A convenient way to do this is using sound waves that can transmit information through variations in pitch, without requiring a line of sight. Vocalizations have developed many times on Earth, in the ocean as well as on land. In fact, whale songs can travel hundreds or even thousands of miles, enabling "verbal" communication between individuals across vast regions of our planet. Hunters tend to be more intelligent than prey animals, because they must employ sophisticated tactics to earn their meal. This is especially true for collaborative pack hunters who coordinate their actions, such as wolves, dolphins, orcas, and chimpanzees.* A solitary human wouldn't stand a chance against

---

* Since intelligence is more likely to arise in hunters but agriculture permits the specialization of skills required for civilization, would extraterrestrials be omnivorous? It's a strong possibility. Flexibility of diet goes hand in hand with intellect, as omnivores must learn to exploit a wide range of food. Some of the most intelligent animals are omnivores, like pigs, bears, badgers, opossums, and raccoons. The next-most intelligent species on Earth, chimpanzees, are omnivorous cooperative hunters like us.

a mammoth, but a group can collaboratively distract, ensnare, and topple the beast. Without some sort of shared language, this would be impossible. Technologically sophisticated beings would want appendages to manipulate tools, and tool use might feed an arms race of increasing brain size and complexity, as it did with humans. For terrestrial animals, having four legs provides support in a bilaterally symmetric system. But find a way to stand up on two feet, and you have two free hands to manipulate tools. So now we have intelligent, talking aliens—with heads, bodies, eyes and ears, two arms and two legs—whose ancestors were hunters.

Or do we? Octopuses are curious and intelligent. They use tools, carrying coconut shells for protection and throwing rocks to break aquarium glass. They have arms to manipulate objects and can be trained to open jars. They can navigate mazes and distinguish between shapes and patterns. Playful, they sometimes release toys into currents, catching them for the sheer fun of it. They've been known to escape their tanks or climb onto fishing boats in search of food, oozing their soft bodies through small cracks. There are well-documented cases of octopuses using their hunting skills to ambush and eat sharks in aquariums at night. Yet, with nine brains and three hearts, octopuses are about as different as we would generally imagine aliens to be.

It's probably because octopuses aren't physically formidable that they developed intelligence, evolving flexible strategies for hunting prey and evading danger. Lacking an arsenal of sharp teeth, octopuses size up their opponents and decide what to do: blending in with a disguise, mimicking a predatory creature, taking shelter, or escaping in a cloud of ink. Arising from the same family as snails and slugs, octopuses surely represent the evolution of a distinct intelligence on Earth. On alien aquatic worlds, we might imagine bands of collaborative octopod hunters founding underwater civilizations.

The truth is that we may not be able to predict what extraterrestrial beings will look like. The variety of life on Earth gives us some clues, and it's easy to speculate why they might be one way or another, but until we're sitting face-to-face with them, the best we can do is guess. What we *can* say is that, whoever and wherever they are, they'll be subject to the same physical laws and will also be carved by billions of years of evolution. We're products of a long chain of events, and they will be, too. But we should never assume that intelligence is a target or goal of evolution. Every organism on Earth, right down to the lowliest bacterium, is at the end of a similarly long chain of improbable events. You've already won the greatest lottery in the universe and are extraordinarily lucky to be alive—but so is a flu virus. For every thousand species ever to exist, only one is still alive today. There've been at least ten distinct species of humans who've walked this planet (members of the genus *Homo**), but only we survived.

Judging from organisms on Earth, a large brain rarely improves evolutionary fitness. Most species spend their energy growing larger teeth, claws, or reproductive organs. While some other Earth creatures display intelligence, human-level intelligence capable of creating technology may be like a peacock's feather, an extravagantly serendipitous outlier. Our ancestors probably evolved intelligence to survive shifting climates in Africa as forests gave way to grasslands. We were forced to walk upright to travel long distances in search of food and to develop more sophisticated hunting tactics for the open plains. Most species wouldn't survive such rapid environmental change, and we nearly didn't. We endured by evolving brains that could invent unique strategies and tools to suit the circumstance. Instead of being optimized to confront one type of challenge, our talent is confronting challenge in general. Humans are specialized generalists, and we might expect the same of extraterrestrial intelligences.

Does intelligence necessarily lead to advanced technology? For millions of years, stone tools were the pinnacle of development. It could easily have ended there, as indeed it did for the majority of human species who didn't make it. For advanced civilizations to emerge on other worlds, intelligent species would have to navigate long chains of discoveries, inventing ways to feed large populations as preconditions for pursuing art, industry, and administra-

---

* The major recognized species are, in alphabetical order: *Homo antecessor, Homo erectus, Homo ergaster, Homo floresiensis, Homo habilis, Homo heidelbergensis, Homo neanderthalensis, Homo rhodensiensis, Homo rudolfensis*, and *Homo sapiens* (us).

tion. Such species would find means of recording and sharing information. They would invent science, mathematics, and engineering to build machines, harness energy, and transport themselves and their products. They would be curious and creative. They would peer into their night skies, trying to understand their place in the cosmos. They would expand across their worlds in pursuit of opportunities and resources. They would be explorers.

But how common would they be? The universe may be filled with advanced civilizations and we just don't know it yet, or we may be utterly alone. But throughout history, the Copernican principle has always borne out.* Once we believed Earth was all there was, but we now know that it's not the center of the universe, the galaxy, or even the solar system. Since we don't know how abundant or rare advanced civilizations might be, it seems safer to assume that there's nothing particularly special about us or our circumstances, that we're not alone. At the conservative end of the Drake Equation, this implies at least fifty other civilizations eventually arising out there in our galaxy. Of course, these wouldn't all develop at once. Since evolutionary processes take billions of years but technological development takes thousands, some civilizations are probably far ahead of us in technology, others so far behind that they don't yet exist.

If they're far ahead of us, we might not be able to detect them because they're using forms of communication we don't yet recognize. But if they're curious, they might keep some old radio detectors tuned in just in case a new primitive civilization like ours comes along, beaming signals into the dark. They'll need powerful radio telescopes to detect our pitifully weak signals, but this shouldn't be an impediment for advanced beings.† They may be too far to have noticed us yet. We've been broadcasting into space for about a century, a mere blink of an eye in the life span of a galaxy. Expanding outward as a bubble, our transmissions have by

---

* Named for the astronomer who proposed that Earth is not the center of the universe, the Copernican principle assumes that there is nothing particularly special about Earth.

† Our radio signals degrade so rapidly that we'd have a tough time detecting ourselves at more than a light-year away.

now reached a tiny region of space encompassing around five hundred stars—less than 0.0001 percent of the galaxy. If there *are* fifty evenly spaced civilizations in our galaxy, the nearest will be a few thousand light-years away, meaning they won't detect our signals for some time. Someone may yet hear our calls, and, given enough time, they may decide to investigate.

It's difficult to predict how an encounter might go. Individuals and cultures of our own species display a wide range of reactions to outsiders: hostility, suspicion, cooperation, sympathy, and idealization.* Throughout the history of humanity, first contact hasn't usually worked out well for the more primitive civilization. However, considering that all advanced beings must successfully navigate a minefield of potential self-annihilation as they grapple with their own technology, there's a good chance they've also outgrown the impulse to go around eradicating more primitive species. This might be the solution to Fermi's Paradox. Perhaps aliens aren't here because they have no reason to be. What motive would drive advanced beings to interrupt the development of civilizations that clearly pose no threat and have little to offer in technological exchange?† Eventually, perhaps, only species who are enlightened enough to survive will be permitted to join the interplanetary congress of worlds. May we be so lucky.

---

* In science fiction, behavioral stereotypes are often applied to entire species. In *Star Trek*, Vulcans are logical, Klingons are violent, Betazoids are emotional, and Ferengi are greedy. No human displays just one of these traits, and we should expect the same diversity from aliens.

† This solution to Fermi's Paradox is unflatteringly known as the "zoo hypothesis." In *Star Trek*, this "Prime Directive" is the driving principle of the United Federation of Planets (which includes Earth).

25 | THE ULTIMATE DESTINATION

We stand today at a crossroads in cosmic history. Billions of years of evolution and thousands of years of technological development have converged to produce a species with the ability to start migrating off its planet. In 1964, astronomer Nikolai Kardashev suggested a scale of civilization development ranging from Type I (harnessing all the energy of its planet) to Type II (harnessing all the energy of its star*) to Type III (harnessing all the energy of its galaxy). Extrapolating on this scale, we currently sit at around Type 0.7, slated to reach Type I in a century or two at our current development rate. But it won't happen automatically. We're going to need fundamental shifts in the way we harness energy, with a complete transition from fossil fuels to renewables and, possibly, widespread fusion power. We'll also have to survive, no small feat in a world of ever-increasing destructive power. As *Star Trek*'s Spock pessimistically observed, "As a matter of cosmic history, it has always been easier to destroy than to create."

The galaxy may be littered with the remains of civilizations that failed to make this transition. If we fail, will alien archaeologists one day come in search of artifacts of long-lost humanity? Will another intelligent species rise on our planet to try again in a

---

* This could be done by building a giant energy-extracting "Dyson sphere" around the star, named after the physicist Freeman Dyson, who proposed the concept in 1960. Astronomers sometimes think they've discovered Dyson spheres around distant stars, but so far none has been confirmed.

hundred million years? Even if we don't wipe ourselves out in a cataclysmic event, progress isn't guaranteed. There've been many instances throughout history of regression, when technology receded. The classical world amassed knowledge that was lost for over a thousand years. Chinese navigators sailed the Indian Ocean and explored the coast of Africa, but the next emperor burned the fleets and the records of these voyages. During the Islamic Golden Age, Baghdad was the global center of scholarship, but a few centuries later the city lay in ashes, its libraries burned and its citizens slain. Even the space program has retreated since the 1960s, its Saturn V Moon rockets dismantled, with many of their blueprints destroyed. Today, we couldn't rebuild one if we wanted to.

Yet, none of these regressions was permanent. We can't estimate the chances of a planet-wide civilization collapse, because our now-global civilization is the only example we know. Are we living in a short window of time where we must travel to space before catastrophe, or can we afford to wait? Will we gradually turn inward, too apathetic, decadent, or good at entertaining ourselves to explore? Will our society be stifled by fake news and pseudoscience? There are more professional astrologers in the United States than astronomers, and almost a quarter of people can't tell the difference. In a time when the network formerly known as the Learning Channel* is showing "documentaries" about alien abductions and ghost sightings, it's no wonder that broad segments of society can't differentiate science and pseudoscience. Nowadays, most people get their information from social media echo chambers, and it's harder than ever to distinguish between reality and nonsense.

On the other hand, people have always predicted the downfall of civilization, but the world is getting better in most measurable ways. Poverty rates are plummeting, literacy has skyrocketed, life span is increasing, wars are disappearing, and crime rates are down. It's a hundred times safer to live in the most dangerous

* Originally founded by NASA and the Department of Education, the Learning Channel was freely broadcast around the planet by satellite. Now privatized as TLC, it no longer makes any pretense about being educational.

American, Chinese, Russian, or even Syrian city than it was to be a prehistoric hunter-gatherer. Our brains are conditioned to disproportionately respond to bad news, and we're bombarded with it. Shootings, famines, earthquakes, and plane crashes sell headlines. The fact that a hundred thousand planes land safely every day doesn't. Even if many people don't understand science, there are more practitioners of it than ever before. In the heyday of ancient Athens, there may have been a thousand scholars. Now there are millions of scientists and engineers working to improve the world, and everyone reaps the rewards. When the swine flu pandemic loomed in 2009, scientists sequenced the virus's genes within a day, and a vaccine was approved within months. Human ingenuity may be able to solve climate change, with green innovations and sustainable technologies. Renewable energy production has been expanding by leaps and bounds. California already produces more than half its electricity from solar during peak sunlight hours. Good news is no reason to be complacent, but it does suggest that we might be in it for the long run.

We don't know if we're alone in the universe, but if we are, that's even more reason to extend the light of consciousness into the cosmos. Once we've spread across many worlds, we'll surely diverge into separate species. Thus, even if there aren't any aliens, the *Star Trek* congress of interstellar beings (all slightly different but vaguely familiar) may become reality—but they would all start human. Divergence could come intentionally through genetic manipulation—perhaps to tailor settlers to the features of individual worlds—or through some kind of selection process. Gradual change driven by natural selection has been the main force of evolution on Earth, but it isn't the only possibility. Simply by selecting for desirable traits throughout history, humans have artificially molded dogs and other domesticated creatures a thousand times more rapidly.

On worlds like Mars with lower gravity, we might grow lighter bones to conserve energy. On worlds bathed in radiation, we might genetically engineer tolerance. We might adapt to breathe different atmospheres and pressures, mimicking, for example, the altitude tolerance possessed by alpine populations on Earth. Natural drift would be incremental, taking thousands of years—partly because sophisticated coping technologies can blunt the "survival advantage" of a given trait. Ten thousand years ago, it would have been catastrophic to have poor eyesight. Today, widely available eyeglasses and contact lenses have eliminated eyesight as a differentiator of who will survive long enough to reproduce. Nevertheless, even without survival pressure, widely scattered humans will drift apart, and many might choose to sculpt themselves to suit the conditions of their new homes.

We'll probably extend the human life span. Obviously, there are personal advantages to living longer, but it also makes sense from an efficiency standpoint. We invest tremendous collective resources into raising new generations, so it's logical to retain their skills as long as possible. Aging is programmed into our genes, but many cells are able to regenerate, so there's no fundamental roadblock to living forever. Since we grew the cells in our body in the first place, we ought to be able to regenerate them indefinitely, and it's actually kind of strange that we can't. Animals such as mole rats, crocodiles, clams, and lobsters don't show any signs of aging. A Greenland shark caught in 2017 was estimated to be 512 years old, born around the time Columbus sailed to America. Our limited life span is simply a product of natural selection, which saw no advantage in keeping us around longer. Progress has already been made at the cellular level: by tweaking certain genes in mice, we can speed or slow their aging. One day, we may be able to program perpetual youth into our cells.

Another path to immortality is to transcend biology, possibly by uploading our consciousness into a computer. We already have very basic brain-computer interfaces, such as cochlear implants, which convert sound waves into nerve signals, allowing deaf patients to hear. Using a neural interface, a paralyzed man kicked the first ball at the 2014 World Cup. With advanced neural interfaces, you could vacation in the rain forest, hear the calls of parrots, and feel the warmth of the Sun, all without leaving your couch. Soon we may be able to cure brain and muscle disorders, either at the neural level or by letting the brain actuate artificial limbs. Scientists have been able to implant artificial memories into mice and transfer memories from one mouse to another. If we can scale this concept, might it be possible to upload memories into the cloud? This could allow the storage of not only thoughts but entire electronic personalities. Re-created in a computer, we could converse with long-lost relatives or famous historical figures.

Entirely transferring our consciousness into a computer is more daunting, especially considering that we're not even sure how consciousness works. But you *are* your brain. If we were to replace one neuron at a time, a hundred billion total, plus all their interconnections, it should be possible to transfer your consciousness into a computer. Freed from your physical body, you could move at the speed of light. Forget *Star Trek* transporters beaming physical bodies; if your neural patterns can be exported, it might be possible in the future to beam your consciousness throughout the galaxy. Protected from extreme temperatures and pressures, you could upload into a robot to explore hellish worlds such as Venus. You could zoom through the clouds of gas giants, touch the Pillars of Creation in the Eagle Nebula, or scale down to the size of a cell and travel through a circulatory system. Even in a nonphysical state, it should be possible to experience the full range of stimuli we can today, plus perhaps others of which we aren't yet aware. You might not even know you were living in a computer.

Or maybe we're *already* living in a simulation? Could the entire universe be no more than a program running on some hyperadvanced alien's laptop? This is a definite possibility. We create simulations all the time, and hyperadvanced aliens millions or billions of years ahead of us in technology ought to be able to do much better. Once you have the ability to create one simulated universe, you can create millions. This could lead to a nesting doll of universes, one inside the other, virtual all the way down—meaning that virtual universes should vastly outnumber real ones. Thus, by the laws of probability, our universe is more likely to be virtual than real. It wouldn't even be necessary to simulate the entire universe, just the region around us, along with spoofed observations of galaxies far out in space. Our bodies and brains could be simulated, responding to simulated stimuli. This is the software version of Fermi's Paradox. Either civilizations never reach the stage where they can make such simulations, they choose not to for some reason, or we're probably living in one.

True, the world feels too real to be simulated. The breeze on your face, the waves crashing on a beach, the aroma of coffee in the morning—all seem perfectly vivid. Perhaps like in *The Matrix*, we biological beings have a peculiar need for tangible reality? But considering that our brains merely interpret electrical impulses, it seems unlikely that we'd be able to tell the difference. Also, since in this scenario we never would have experienced "reality" in the first place, we wouldn't even know what it was. The only way to tell simulation from reality is to identify flaws in the program, perhaps by finding inconsistencies in the laws of physics. In fact, the laws of physics do seem suspiciously designed. Gravity, electromagnetism, and nuclear forces are determined by constants that fall into extremely narrow ranges capable of supporting stars, planets, life, and, well, us. Tweak any of them up or down, and everything falls apart. But this doesn't necessarily imply the universe was designed. It could be that there are many parallel universes, but we live in this one because it's the only one we *can* live in. Or it could be that there's some unknown reason the laws of physics can only be the way they are.

What does the future hold? If the past has taught us anything, it's that the future will be different from anything we can reasonably predict. Technologies of the far future, whatever they may be, will seem like magic to us, according to Clarke's Third Law.* Airplanes seem normal—even mundane—to us, but they would be incomprehensible to our distant ancestors. It's actually kind of sad that we can find something as miraculous as flying routine. The ancient Greeks imagined only their gods capable of such feats. We become conditioned to find the wondrous boring and only the novel fascinating, which might mean that immortality, and the opportunity it presents to have all our questions answered, would

---

* Arthur C. Clarke's laws: 1) When a distinguished but elderly scientist states that something is possible, he is almost certainly right. When he states that something is impossible, he is very probably wrong. 2) The only way of discovering the limits of the possible is to venture a little way past them into the impossible. 3) Any sufficiently advanced technology is indistinguishable from magic.

be a curse. Oh, to be a child again, naturally curious, unafraid to ask questions, and swimming in things we don't know.

But the universe's proportions and constant expansion ensure that there'll always be puzzles to solve. Since none of us—not even the aliens we've yet to meet—can ever reach the universe's boundary, there will always be something to learn, some new wonder to explore.

Could we lose our exploratory zeal and become too comfortable at home? With our culture of smartphones, 3-D movies, and social media, do we risk becoming the Eloi from *The Time Machine*, living banal lives of ease? Not if we remember that a bit of discomfort can be a good thing—and so can a bit of risk. Life is, after all, a one-way journey, with an expected survival chance of precisely zero. Against this terrifying calculus, striving for absolute safety at the expense of accomplishment is certain to be a bad bargain. Exploration is worth the risk because it brings meaning to life.

After billions of years of evolution, we're finally ready to leave our planet and see the universe for ourselves. If we choose to take up this challenge, there's little doubt that from the hindsight of history, it will be one of the most important decisions we've ever made. Centuries ago, if you traveled the world, you'd embark on a rickety sailing ship and spend months in cramped, dark, and damp quarters. You might never go above decks to get a breath of fresh air. Along the way, you'd face a daunting array of hazards: pirates and war, storms and shipwreck, malnutrition and disease. This was the routine, along well-charted routes. Voyages into the unknown were far more perilous still, and many who set out never returned. Our ancestors accepted these risks, and prospered as a result.

The challenges we face today are no more daunting than they've been in the past, and the rewards no less meaningful. Pushing our boundaries is the best way to expand them, and to unite humanity in common purpose. Will we turn from this calling and abandon exploration? I don't think so. There will always be wanderers among us.

EPILOGUE

June 28, 2015, was a Sunday. Most of us SpaceXers in Hawthorne, California, had come in to watch the Dragon spacecraft launch to the International Space Station, which is pretty typical even for a weekend. The time of a launch is mostly based on orbital considerations (you have to launch to the right place in orbit), and this was a morning launch. The crowds were larger than usual because everyone was hoping for the first successful booster landing. The previous landings had come so close to working. We were learning with each attempt, crossing off anything that had gone wrong. My friend had brought his three-year-old son and another child he was looking after to watch the launch. Everyone was expectant, excited.

A few of us were clustered around some computers upstairs watching data come in. Five, four, three, two, one . . . first-stage engines lit! One and a half million pounds of thrust took Falcon 9 into the sky. Telemetry was normal; everything was going smoothly. Then, 139 seconds into flight, as I watched the video monitor, there was a sudden bright flash and a giant puff of smoke. No rocket emerged from the other side.

Normally, as a rocket enters the upper atmosphere, the engine plume mushrooms out into the thin air and generates a lot of smoke. If you're looking closely, this can be alarming the first time you see it, particularly at night, when the exhaust plume shines bright and the smoke seems to glow. For a second or two, I thought this effect was what I was looking at. I thought: *Maybe the camera is lingering on the engine smoke for longer than usual.*

Then realization crept over us, that horrible sinking feeling when you wish you could just rewind the clock. "We lost it," my friend said in disbelief. The kids watching the launch were too young to understand what had just happened, but seeing the looks on our faces, they probably sensed that something was wrong. My friend jumped up and said, "Okay, ice-cream time!" and then took them on an unscheduled trip to Disney. Theme parks work wonders in erasing rocket failures from the minds of toddlers.

Everyone else started combing through data in an extremely calm and professional way. Any suspicious readings were flagged and added to a growing tree of possible causes. Every possible contributor was investigated and traced to its source. After a thorough investigation, the cause was found and safeguards put in place, and a mere six months later, on December 21, 2015, Falcon 9 was ready to fly again.

Not only was the launch a complete success, but for the first time in history, the orbital-class rocket booster landed successfully on that flight. Now, dozens of flights later, landing the booster is the norm rather than the exception. On March 30, 2017, a recovered rocket flew to space a *second* time, and successfully landed once again. Reusable rockets are a technological revolution with the promise to vastly expand access to space, but breakthroughs like this wouldn't be possible without pushing limits. If you're not breaking stuff on occasion, you're not trying hard enough.

If we devote too much effort to avoiding failure, we're not devoting enough to achieving success. This doesn't mean we should be reckless; rather, it means that we should take calculated risks at the edge of our technological abilities. Choosing not to take risks merely guarantees a different kind of failure. If the history of exploration has taught us anything, it's that amazing things happen when humans force themselves to try something no one has ever done before. If our ancestors hadn't been willing to take risks, humans would still today be an interesting but insignificant species confined to the African Rift Valley. But we come from a line of explorers, people who resolutely challenged the impossible to create a future for themselves and their descendants. We're capable of incredible things, if we're only willing to try.

ACKNOWLEDGMENTS

I've always been fascinated by history, but I didn't always see today's incremental progress in space exploration as an integral component of the human future. This realization emerged in my college years, propelled by the persistent persuasion of my friend Iain, the most inspirational soul I've ever encountered. He taught me that we can't just sit around and wait for the future—we have to create it. I'm also deeply indebted for my early inspiration to Roger Taguchi, a teacher whose boundless excitement about the world (and love of trivia) is irresistibly infectious. Several university professors stand out, in particular my master's advisor Fred Afagh and Ph.D. advisors Chuck Oman and Dan Merfeld.

This book wouldn't have been possible without a chance encounter on an airplane; I will always be grateful to Kristin Loberg and Bonnie Solow for taking me under their wings. I'm also greatly indebted to Rick Horgan and the Scribner publishing family, in particular, Nan Graham, Colin Harrison, Roz Lippel, Emily Greenwald, Paul O'Halloran, Ashley Gilliam, Jaya Miceli, Rosie Mahorter, Brian Belfiglio, and Jason Chappell (along with the Canadian team, Mackenzie Croft and Jessica Scott). Aja Pollock, Jennifer Racusin, Annarosa Schiavone, Julian Smith, John Reynen, and Peter Cawdron did wonderful jobs providing thorough edits and invaluable feedback. Finally, I'd like to express my profound appreciation to my wife Lisa for her tireless chapter reviews, devoted support, and enduring patience with late evenings and long weekends of writing.

NOTES ON SOURCES

PART I: IN THE BEGINNING

CHAPTER 1: OUT OF THE CRADLE

The human diaspora from Africa is somewhat murky because we have only fragmentary evidence for many specimens. For example, for *Homo antecessor*, one of the first post–*Homo erectus* species to leave Africa, we have incomplete fossils from only six individuals. Some authors suggest that *Homo antecessor* was in the direct lineage of both *Homo sapiens* and *Homo Neanderthalensis* (e.g., José-Maria Bermúdez de Castro et al., "A Hominid from the Lower Pleistocene of Atapuerca, Spain: Possible Ancestor to Neanderthals and Modern Humans," *Science* 276, no. 5317 [1997]:1392–95), but others propose that *Homo antecessor* represents a dead-end branch (e.g., Richard Klein, "Chapter 3: Hominin Dispersals in the Old World," in *The Human Past: World Prehistory and the Development of Human Societies*, second edition, edited by Chris Scarre, [London: Thames & Hudson, 2009], 84–123). Most archaeologists agree that *Homo heidelbergensis* was an ancestor of both modern humans in Africa as well as Neanderthals in Eurasia (Luigi Luca Cavalli-Sforza and Francesco Cavalli-Sforza, *The Great Human Diasporas*, translated by Sarah Thorne [New York: Perseus Books, 1996]). For the casual reader, the Smithsonian Museum of Natural History includes a good summary on its website (www.humanorigins.si.edu/evidence/human-fossils/species/).

The Toba supervolcano eruption seventy-five thousand years ago is well substantiated (e.g., Stanley Ambrose, "Late Pleistocene Human Population Bottlenecks, Volcanic Winter, and Differentiation of Modern Humans," *Journal of Human Evolution* 34, no. 6 [1998]: 623–51), but the suggestion that it created a human population bottleneck is disputed (e.g., Christine Lane et al., "Ash from the Toba Supereruption in Lake Malawi Shows No

Volcanic Winter in East Africa at 75 ka," *Proceedings of the National Academy of Sciences* 110, no. 20 [2013]: 8025–29).

The emergence of modern humans in Africa is well documented (Christopher S. Henshilwood et al., "A 100,000-Year-Old Ochre-Processing Workshop at Blombos Cave, South Africa," *Science* 334, no. 6053 [2011]: 219–22), as is the fact that humans share significant portions of their DNA with Neanderthals (Kay Prüfer et al., "A High-Coverage Neandertal Genome from Vindija Cave in Croatia," *Science* 358, no. 6363 [2017]: 655–58). Discoveries of rock-and-mammoth-bone settlements at Dolní Věstonice are summarized in Gene Stuart, "Ice Age Hunters: Artists in Hidden Cages," *Mysteries of the Ancient World* (Washington, DC: National Geographic Society, 1979).

CHAPTER 2: EARLY WANDERINGS

Jared Diamond's *Guns, Germs, and Steel* (New York: W. W. Norton, 1997) touches on the development of Aboriginal fishing communities. A summary of archeological evidence can be found on the Australian Department of the Environment and Energy's website (www.environment.gov.au/heritage/places/national/budj-bim/).

Thickness of the continental ice sheets is described in Richard Foster Flint's *Glacial and Quaternary Geology* (New York: John Wiley, 1971). A history of Inuit arrival from Asia and dispersal in North America can be found in *The Cambridge History of the Native Peoples of the Americas: Volume 1*, edited by Bruce G. Trigger and Wilcomb E. Washburn (Cambridge: Cambridge University Press, 1996). Competition between the Inuit and Vikings is vividly described by Jared Diamond in *Collapse: How Societies Choose to Fail or Succeed* (New York: Viking, 2006).

Estimates of the population of the Americas before European arrival range drastically—from less than ten million to more than a hundred million. Charles C. Mann convincingly argues that it was on the higher end of this range in *1491: New Revelations of the Americas Before Columbus* (New York: Knopf, 2005). The fact that syphilis was present in the Americas before European contact is documented in Bruce Rothschild et al., "First European Exposure to Syphilis: The Dominican Republic at the Time of Columbian Contact," *Clinical Infectious Diseases* 31, no. 4 (2000): 936–41.

CHAPTER 3: PEOPLE OF THE SEA

DNA evidence and timelines for the peopling of Polynesia are recounted in Pontus Skoglund et al., "Genomic Insights into the Peopling of the Southwest Pacific," *Nature* 538 (2016): 510–13.

NOTES ON SOURCES

A fascinating account of Polynesian navigational methods, and on Tupaia's ability to recount details about islands that he'd never visited, can be found in Joan Druett's *The Remarkable Story of Captain Cook's Polynesian Navigator* (New Jersey: Old Salt Press, 1987). For good summaries of Polynesian navigation, also see Jared Diamond's *Collapse: How Societies Choose to Fail or Succeed* (New York: Viking, 2006) and Geoffrey Irwin's *The Prehistoric Exploration and Colonization of the Pacific* (Cambridge: Cambridge University Press, 1992).

The linguistic connection between Polynesian islands and the encounter between Tupaia and the Maori are described in Michael King's *The Penguin History of New Zealand* (Auckland: Penguin Books, 2003).

Evidence of Polynesian southward ventures toward Antarctica is documented in Atholl Anderson and Gerard O'Regan, "To the Final Shore: Prehistoric Colonisation of the Subantarctic Islands in South Polynesia," *Australian Archaeologist: Collected Papers in Honour of Jim Allen*, edited by Atholl Anderson and Tim Murray (Canberra: Coombs Academic Publishing, Australian National University, 2000), 440–54, and on the Australian Antarctic Division website (www.antarctica.gov.au/about-antarctica /history/stations/macquarie-island).

CHAPTER 4: ANTIQUITY

Archeological evidence for stone tools on Crete dating back 130,000 years is summarized in Andrew Lawler, "Neandertals, Stone Age People May Have Voyaged the Mediterranean," *Science*, online edition, http:// dx.doi.org/10.1126/science.aat9795 (2018).

DNA evidence for mummified baboons brought back from Punt to Egypt was presented in a podium session by Nathaniel J. Dominy et al. at the 84th Annual Meeting of the American Association of Physical Anthropologists, titled "Mummified Baboons Clarify Ancient Red Sea Trade Routes," an abstract of which can be found in the AAPA Presentation Schedule, *American Journal of Physical Anthropology* 156 (2015): 122–23. The description of *The Tale of the Shipwrecked Sailor* is based on Vejas Gabriel Liulevicius's excellent lecture series *History's Greatest Voyages of Exploration*, "Lecture 1: The Earliest Explorers," (Chantilly, VA: The Great Courses, 2015). The case for Egyptian forays into the African interior is convincingly argued by Paul Herrmann in *The Great Age of Discovery*, translated by Arnold J. Pomerans (New York: Harper, 1958; reprinted by Whitefish, MT: Kessinger, 2007). Evidence for the Egyptian-okapi connection can be found in Captain C. H. Stigand, "The Lost Forests of Africa," *Geographical Journal* 45 (London: The Royal Geographical Society, 1915), 513–20.

The voyages of Hanno the Navigator are based on a translated account with commentary by Jona Lendering, "Hanno the Navigator," Livius, https://www.livius.org/articles/person/hanno-1-the-navigator/ (last modified April 21, 2019).

The description of the Great Library of Alexandria and the scientific contributions of Eratosthenes were inspired by Carl Sagan's masterpiece televised series (*Cosmos: A Personal Voyage*, episode 1, "The Shores of the Cosmic Ocean," aired September 28, 1980, on PBS) and accompanying book (*Cosmos* [New York: Random House, 2002]).

CHAPTER 5: THE CLASSICAL WORLD

The assertion that the Persian Empire contained almost half the people in the world is based on Rein Taagepera, "Size and Duration of Empires: Growth-Decline Curves, 600 B.C. to 600 A.D.," *Social Science History* 3, no. 3/4 (1979): 115–38. The account of the emergence of Athens and the splendor of Greek culture were partly inspired by Eric Weiner's *The Geography of Genius: Lessons from the World's Most Creative Places* (New York: Simon & Schuster, 2016), along with Carl Sagan's *Cosmos* (New York: Random House, 2002). Susan Wise Bauer's *The History of the Ancient World: From the Earliest Accounts to the Fall of Rome* (New York: W. W. Norton, 2007) was an indispensable reference for details. Also see Herodotus, translated by A. D. Godley, *The Persian Wars, Volume 1: Books 1–2*, book 117 in the Loeb Classical Library series (Cambridge, MA: Harvard University Press, 1920).

For further reading on the travels of Alexander the Great, see Philip Freeman's wonderful account *Alexander the Great* (New York: Simon & Schuster, 2011). Details of Pytheas's travels is based on Barry Cunliffe's *The Extraordinary Voyage of Pytheas the Greek: The Man Who Discovered Britain*, revised edition (New York: Walker, 2002); a good summary and the tidbit about Britons asserting the Irish were cannibals are included in Vejas Gabriel Liulevicius's *History's Greatest Voyages of Exploration*, "Lecture 2: The Scientific Voyage of Pytheas the Greek" (Chantilly, VA: The Great Courses, 2015).

The account of the Roman Empire was inspired by Mary Beard's excellent book *SPQR: A History of Ancient Rome* (New York: Liveright, 2015). The description of Roman naval technological development, in particular the reverse-engineering of a Carthaginian vessel, is described in J. F. Lazenby's *The First Punic War: A Military History* (Stanford, CA: Stanford University Press, 1996). Roman trade with India and a comparison of the Roman pepper trade with Venetian and Portuguese imports are documented in G. V. Scammell's *The World Encompassed: The First European Maritime Empires c. 800–1650* (Berkeley and Los Angeles: University of California Press, 1981).

The theory that Roman soldiers made their way to China following the Battle of Carrhae was suggested by H. H. Dubs, "A Roman City in Ancient China," *Greece and Rome* 4, no. 2 (1957): 139–48. The idea is fascinating (and somewhat plausible), yet any theory based on a convoluted chain of events so long ago should be approached with a great deal of skepticism. For more discussion, see Erling Hoh, "Lost Legion," *Far Eastern Economic Review* (January 14, 1999): 60–62.

PART II: REDISCOVERING THE WORLD

CHAPTER 6: BARBARIANS FROM THE NORTH

For a vivid and surprisingly modern-sounding account of the people and technology of pre-Roman Northern Europe, see Julius Caesar's *The Gallic Wars*, translated by W. A. McDevitte and W. S. Bohn, hosted on the Massachusetts Institute of Technology's classics website (http://classics .mit.edu/Caesar/gallic.html); the naval battle with the Veneti is described in book 3, chapter 13.

Roman military forays into Ireland and the Irish role in preserving classical works are described in Vittorio Di Martino's *Roman Ireland* (Doughcloyne, Ireland: The Collins Press, 2006) and Thomas Cahill's *How the Irish Saved Civilization: The Untold Story of Ireland's Heroic Role from the Fall of Rome to the Rise of Medieval Europe* (New York: Doubleday, 1995), respectively.

The account of Irish monks and the voyages of St. Brendan are based on T. C. Lethbridge's *Herdsmen and Hermits: Celtic Seafarers in the Northern Seas* (Cambridge: Bowes & Bowes, 1950) and Vejas Gabriel Liulevicius's *History's Greatest Voyages of Exploration*, "Lecture 3: St. Brendan: The Travels of an Irish Monk" (Chantilly, VA: The Great Courses, 2015). The assertion that Columbus sailed to the Americas with a map that showed an "Island of St. Brendan" is from *The Classical Tradition and the Americas*, edited by Wolfgang Haase and Reinhold Meyer (Berlin: Walter de Gruyter, 1993).

Description of Viking society and the origin of the word "berserk" were inspired by Liulevicius's "Lecture 5: Leif Eriksson the Lucky." Viking consumption of sauerkraut, onions, and radishes as a defense against scurvy is recounted in Paul Herrmann's *The Great Age of Discovery*, translated by Arnold J. Pomerans (New York: Harper, 1958; reprinted by Whitefish, MT: Kessinger, 2007). Evidence for Viking use of a polarized mineral to focus light on cloudy days is reviewed by Albert Le Floch et al., "The Sixteenth Century Alderney Crystal: A Calcite as an Efficient Reference Optical Compass?," *Proceedings of the Royal Society A: Mathematical, Physical and Engineering Sciences* 469, no. 2153 (2013), https://doi.org/10.1098/rspa.2012.0651.

Descriptions of Viking lifestyle and expansion owes much to Jared Diamond's *Collapse: How Societies Choose to Fail or Succeed* (New York: Viking, 2006). Details of forays into North America draw on "The Saga of Erik the Red," translated by J. Sephton in 1880, Icelandic Saga Database, sagadb.org/eiriks_saga_rauda.en (accessed April 24, 2019). Evidence for butternut found in Newfoundland is presented in Birgitta Wallace, "The Norse in Newfoundland: L'Anse aux Meadows and Vinland," special issue, *New Early Modern Newfoundland: Part 2* 19, no. 1 (2003). The eleventh-century Norwegian penny is discussed in Patricia Sutherland, "The Norse and Native Norse Americans," *Vikings: The North Atlantic Saga*, edited by William F. Fitzhugh and Elizabeth Ward (Washington, DC: Smithsonian Books, 2000). Radiocarbon dating of Greenland Viking artifacts to as late as 1450 is asserted in Jette Arneborg et al., "Change of Diet of the Greenland Vikings Determined from Stable Carbon Isotope Analysis and ^{14}C Dating of Their Bones," *Radiocarbon* 41, no. 2 (1999): 157–68.

CHAPTER 7: EARLY CONTACTS

The case for early contacts owes much to the arguments presented in Paul Herrmann's *The Great Age of Discovery*, translated by Arnold J. Pomerans (New York: Harper, 1958; reprinted by Whitefish, MT: Kessinger, 2007); however, significant new evidence has surfaced since that book was originally published, strengthening some of the book's arguments while weakening others. In particular, evidence for contact between Polynesia and North America based on the sweet potato seems to be accumulating, based on a number of sources, including Brian Switek, "DNA Shows How the Sweet Potato Crossed the Sea," *Nature News* (2013), http://dx.doi.org/10.1038/nature.2013.12257; Caroline Roullier et al., "Historical Collections Reveal Patterns of Diffusion of Sweet Potato in Oceania Obscured by Modern Plant Movements and Recombination," *Proceedings of the National Academy of Sciences* 110, no. 6 (2013): 2205–10; and Álvaro Montenegro et al., "Modeling the Prehistoric Arrival of the Sweet Potato in Polynesia," *Journal of Archaeological Science* 35, no. 2 (2008): 355–67.

Evidence for pre-Columbian contact based on coconut dispersal has also been recently strengthened by Luc Baudouin et al., "The Presence of Coconut in Southern Panama in Pre-Columbian Times: Clearing Up the Confusion," *Annals of Botany* 113, no. 1 (2014): 1–5, and John L. Sorenson and Carl L. Johannessen, "Scientific Evidence for Pre-Columbian Transoceanic Voyages," *Sino-Platonic Papers* 133 (2004).

The observations by Adelbert von Chamisso on South American naval technology are recounted in Isabel Ollivier's review of *A Voyage Around the*

World with the Romazov Exploring Expedition in the Years 1815–1818 in the Brig Rurik in *Journal of the Polynesian Society* 98, no. 1 (1989): 97–99. Discussion on pre-Columbian South American voyages and raft design draw on Jeff Emanuel, "Crown Jewel of the Fleet: Design, Construction, and Use of the Seagoing Balsas of the Pre-Columbian Andean Coast," *Proceedings of the 13th International Symposium on Boats and Ship Archaeology* (October 2012) and Thor Heyerdahl and Arne Skjölsvold, "Archaeological Evidence of Pre-Spanish Visits to the Galápagos Islands," *Memoirs of the Society for American Archaeology* 12 (1956).

For more on canoe and tool design similarities, see José-Miguel Ramírez-Aliaga, "The Polynesian-Mapuche Connection: Soft and Hard Evidence and New Ideas," *Rapa Nui Journal* 24, no. 1 (2010): 29–33, and Terry L. Jones and Kathryn A. Klar, "Diffusionism Reconsidered: Linguistic and Archaeological Evidence for Prehistoric Polynesian Contact with Southern California," *American Antiquity* 70, no. 3 (2005): 457–84. Linguistic ties are outlined in Willem F. H. Adelaar with Pieter C. Muysken, "Genetic Relations of South American Indian Languages," *Languages of the Andes*, 41 (Cambridge: Cambridge University Press, 2004), and Simon Greenhill et al., "Entries for KUMALA.1 [LO] Sweet Potato (Ipomoea)," *POLLEX-Online: The Polynesian Lexicon Project Online* (2010). The case for contact based on chicken DNA is described in Vicki A. Thompson et al., "Using Ancient DNA to Study the Origins and Dispersal of Ancestral Polynesian Chickens Across the Pacific," *Proceedings of the National Academy of Sciences* 111, no. 13 (2014): 4826–31.

The modern interpretation on the "white god" myth (originally recounted by Herrmann) was partly inspired by Camilla Townsend, "Burying the White Gods: New Perspectives on the Conquest of Mexico," *American Historical Review* 108, no. 3 (2003): 659–87.

CHAPTER 8: THE OTHER MEDITERRANEAN

The Portuguese forays into the Indian Ocean, along with King Manuel I's quip "It is not we who have discovered them, but they who have discovered us," is discussed in Anthony Pagden's *Peoples and Empires: A Short History of European Migration, Exploration, and Conquest, from Greece to the Present* (New York: Modern Library, 2001).

Details on the Majapahit Empire and Arabs sailors of the Indian Ocean are outlined in Bertold Spuler's *The Muslim World: A Historical Survey* (Brill Archive, 1981), and the embassy of Odoric of Pordenone is discussed in Georges Maspero's *The Champa Kingdom: The History of an Extinct Vietnamese Culture*, translated by E. J. Tips (Banglamung, Thailand: White Lotus Press, 2002).

The account of the travels of Ibn Battuta are based on Ross E. Dunn's *The Adventures of Ibn Battuta: A Msulim Traveler of the Fourteenth Century* (Berkeley and Los Angeles: University of California Press, 2005) and the excellent summary by Vejas Gabriel Liulevicius in *History's Greatest Voyages of Exploration*, "Lecture 7: Ibn Battuta: Never the Same Route Twice" (Chantilly, VA: The Great Courses, 2015).

Vasco da Gama's journals from his voyage to India (*A Journal of the First Voyage of Vasco da Gama, 1497–1499*, translated by E. G. Ravenstein [New Delhi: Asian Educational Services, 1995]), are a fascinating read and include tidbits such as the fact that he brought along Jewish interpreters who tried to speak Arabic to sailors along the Swahili coast.

An analysis of the spice trade, including the relative competitiveness of Venice and Portugal and how salt and pepper came to stock our dinner tables appears in the excellent lecture series on culinary history by Ken Albala, *Food: A Cultural Culinary History* (Chantilly, VA: The Great Courses, 2013). The reference to Alaric the Goth demanding peppercorns is from *Black Pepper:* Piper nigrum, edited by P. N. Ravindram (Boca Raton, FL: CRC Press, 2000).

The discussion of Ottoman naval power and its challenge to Portuguese dominance of the Indian Ocean draws upon Giancarlo Casale's *The Ottoman Age of Exploration* (New York: Oxford University Press, 2010).

CHAPTER 9: CHINA'S AGE OF EXPLORATION

The role China played in the development of naval technology is described by Joseph Needham in *Science and Civilization in China*, volume 4, *Physics and Physical Technology* (Taipei: Caves Books, 1986), as well as by David Graff and Robin Higham in *A Military History of China* (Boulder, CO: Westview Press, 2002).

Description of the Mongol conquests and the role they played in diffusing technologies is based on the wonderful account by Jack Weatherford in *Genghis Khan and the Making of the Modern World* (New York: Crown, 2004). Recommended reading on the topic also includes S. Frederick Starr's *Lost Enlightenment: Central Asia's Golden Age from the Arab Conquest to Tamerlane* (Princeton, NJ: Princeton University Press, 2013), though the scope of this work is far broader.

The summary of Marco Polo's journey is based on Henry Hersch Hart's *Marco Polo, Venetian Adventurer* (Norman: University of Oklahoma Press, 1967), and also on Vejas Gabriel Liulevicius's *History's Greatest Voyages of Exploration*, "Lecture 6: Marco Polo and Sir John Mandeville" (Chantilly, VA: The Great Courses, 2015); he characterized Marco Polo as "the first modern secular humanist."

The contrast between China and Europe, and the fact that in 1400, Europe accounted for a mere 10 percent of the world's land surface and 15 percent of its population is based on Niall Ferguson's *Civilization: The West and the Rest* (New York: Penguin Books, 2012). Evidence for beer in ancient China is presented in Jiajing Wang et al., "Revealing a 5,000-Y-Old Beer Recipe in China," *Proceedings of the National Academy of Sciences* 113, no. 23 (2016): 6444–48.

Recommended readings on the voyages of Zheng He and the Chinese treasure fleets include Edward L. Dreyer's *Zheng He: China and the Oceans in the Early Ming, 1405–1433* (Old Tappan, NJ: Pearson Longman, 2006) and Louise Levathes's *When China Ruled the Seas: The Treasure Fleet of the Dragon Throne, 1405–1433* (New York: Oxford University Press, 1996).

CHAPTER 10: A SEA ROUTE TO INDIA

The discussion on Portugal's founding and naming is based on Paul Herrmann's *The Great Age of Discovery*, translated by Arnold J. Pomerans (New York: Harper, 1958; reprinted by Whitefish, MT: Kessinger, 2007). Details on Henry the Navigator and the overseas expansion of Portugal draw on Bailey W. Diffie and George D. Winius's *Foundations of the Portuguese Empire, 1415–1580* (Minneapolis: University of Minnesota Press, 1977). The assertion that at least thirty shipwrecks occurred between 1790 and 1806 near Cape Bogador is found in Dean H. King's *Skeletons on the Zahara: A True Story of Survival* (New York: Little, Brown, 2004).

The account of Vasco da Gama's voyage to India is based on his journals (*A Journal of the First Voyage of Vasco da Gama, 1497–1499*, translated by E. G. Ravenstein [New Delhi: Asian Educational Services, 1995]), Herrmann's *Great Age of Discovery*, and the excellent summary by Vejas Gabriel Liulevicius in *History's Greatest Voyages of Exploration*, "Lecture 8: Portugal's Great Leap Forward" (Chantilly, VA: The Great Courses, 2015).

Also recommended is John Jones's 1863 translation of *The Itinerary of Ludovico di Varthema of Bologna from 1502 to 1508* (New Delhi: Asian Educational Services, 1997), which gives a fascinating account of a European adventurer in Asia.

CHAPTER 11: PLUNDER AND GOLD

The summary of Columbus's life and travels draw on Paul Herrmann's *The Great Age of Discovery*, translated by Arnold J. Pomerans (New York: Harper, 1958; reprinted by Whitefish, MT: Kessinger, 2007) and

Vejas Gabriel Liulevicius's *History's Greatest Voyages of Exploration* "Lecture 9: The Enigmatic Christopher Columbus" (Chantilly, VA: The Great Courses, 2015).

The discussion on the humanist efforts of Bartolomé de Las Casas and his contrast with Columbus was inspired by David M. Traboulay's *Columbus and Las Casas: The Conquest and Christianization of America, 1492–1566* (Lanham, MD: University Press of America, 1994).

Discussions on the state of Native American societies before European arrival and the conquests of the conquistadors draw from Jared Diamond's *Guns, Germs, and Steel* (New York: W. W. Norton, 1997) and Charles C. Mann's *1491: New Revelations of the Americas Before Columbus* (New York: Knopf, 2005). In *1491*, Mann convincingly argues for population estimates on the high side for Native American societies (up to sixty million in Mexico) and suggests that twenty thousand people were sacrificed by the Aztecs every year. His *1493: Uncovering the New World Columbus Created* (New York: Knopf, 2011) is a fabulous investigation into the impacts of Columbus's voyages; this book supplied general intellectual fodder for "Plunder and Gold" and influenced other chapters, too.

CHAPTER 12: AROUND THE WORLD

The roles played by Amerigo Vespucci and Martin Waldseemüller in the naming of America are discussed in John Wilford's *The Mapmakers: The Story of the Great Pioneers in Cartography—from Antiquity to the Space Age*, revised edition, (New York: Knopf, 2000).

The summary of Magellan's voyage in this chapter draws mainly on Laurence Bergreen's gripping book *Over the Edge of the World: Magellan's Terrifying Circumnavigation of the Globe* (New York: William Morrow, 2003). A few details (e.g., the fact that many of Magellan's crew were scoured from prison) are from Paul Herrmann's *The Great Age of Discovery*, translated by Arnold J. Pomerans (New York: Harper, 1958; reprinted by Whitefish, MT: Kessinger, 2007). Vejas Gabriel Liulevicius's *History's Greatest Voyages of Exploration*, "Lecture 10: Magellan and the Advent of Globalization" (Chantilly, VA: The Great Courses, 2015) was also consulted.

For further reading on Spanish Manila and the Acapulco galleon route, see J. H. Parry's *The Spanish Seaborne Empire* (New York: Knopf, 1967) and Charles C. Mann's *1493: Uncovering the New World Columbus Created* (New York: Knopf, 2011).

PART III: MODERNITY

CHAPTER 13: EMPIRES OF TRADE

The discussion in this chapter on the rise of Europe owes much to Niall Ferguson's *Civilization: The West and the Rest* (New York: Penguin Books, 2012). The connection between the rise of Europe and exploration is examined (though mostly indirectly) in Anthony Pagden's *The Enlightenment: And Why It Still Matters* (New York: Random House, 2013). Angus Maddison's *Growth and Interaction in the World Economy: The Roots of Modernity* (Washington, DC: AEI Press, 2001) was also consulted for its description of the rise of European economic prosperity. Details on literacy rates and the influence of the printing press can be found in Robert Houston's *Literacy in Early Modern Europe: Culture and Education, 1500–1800*, second edition (Chicago: Addison-Wesley Longman, 1989) and Elizabeth L. Eisenstein's *The Printing Revolution in Early Modern Europe*, Canto Classics, second edition (Cambridge: Cambridge University Press, 2012).

For further reading on the rise of the Netherlands and its global fleet, see Jonathan Israel's *The Dutch Republic: Its Rise, Greatness and Fall, 1477–1806* (New York: Oxford University Press, 1995). The assertion that more than sixteen thousand Dutch ships plied the world's oceans by 1650 is made in Steven Mintz and Sara McNeil's "The Middle Colonies: New York," *Digital History* (http://www.digitalhistory.uh.edu/disp_textbook.cfm?smtID=2&psid=3586).

The discussion of the influence of Dutch New Amsterdam is based on Russell Shorto's excellent book *The Island at the Center of the World: The Epic Story of Dutch Manhattan and the Forgotten Colony That Shaped America* (New York: Vintage, 2005). Description of the contrast between the geographic expanse and economic prosperity of the French and English colonies was inspired by Niall Ferguson's *The Ascent of Money: A Financial History of the World* (New York: Penguin Books, 2008).

For further reading on free maroon societies in the Americas, a good starting point is Charles C. Mann's excellent account in *1493:Uncovering the New World Columbus Created* (New York: Knopf, 2011).

CHAPTER 14: OPENING CONTINENTS

For further reading on Henry Hudson's voyage, see Doug Hunter's *God's Mercies: Rivalry, Betrayal, and the Dream of Discovery* (Toronto: Doubleday Canada, 2007). For more on Alexander Mackenzie's tracing of his eponymous river into the Arctic, see Barry Gough's *First Across the Continent: Sir Alexander Mackenzie* (Norman: University of Oklahoma Press, 1997).

The description of the Lewis and Clark expedition in this chapter draws heavily on Stephen E. Ambrose's masterpiece account of the Corps of Discovery, *Undaunted Courage: Meriwether Lewis, Thomas Jefferson, and the Opening of the American West* (New York: Simon & Schuster, 1997). Vejas Gabriel Liulevicius's *History's Greatest Voyages of Exploration*, "Lecture 16: Jefferson Dispatches Lewis and Clark" (Chantilly, VA: The Great Courses, 2015) was also consulted, contributing the tidbit about mammoths being considered as the US national animal in place of the bald eagle.

The summary of Russian expansion, and in particular the incredible voyage of Vitus Bering, was inspired by a reading of Orcutt Frost's *Bering: The Russian Discovery of America* (New Haven, CT: Yale University Press, 2003).

CHAPTER 15: FRONTIERS OF SCIENCE

For further reading on John Harrison's method of computing longitude, see Dava Sobel's wonderful book *Longitude: The True Story of a Lone Genius Who Solved the Greatest Scientific Problem of His Time* (London: Walker, 1995). Brian Richardson's *Longitude and Empire: How Captain Cook's Voyages Changed the World* (Vancouver: University of British Columbia Press, 2005) expands on how the discovery of a method of computing longitude enabled Cook's voyages.

The descriptions of Captain Cook's voyages draw on Peter Aughton's *Endeavour: The Story of Captain Cook's First Great Epic Voyage* (Glouchestershire, UK: Windrush Press, 1999), Philip Edwards's *James Cook: The Journals* (London: Penguin Books, 2003), and Paul Herrmann's *The Great Age of Discovery*, translated by Arnold J. Pomerans (New York: Harper, 1958; reprinted by Whitefish, MT: Kessinger, 2007).

The account of Alexander von Humboldt's travels was inspired by Andrea Wulf's *The Invention of Nature: Alexander von Humboldt's New World* (New York: Knopf, 2015). Humboldt's characterization of himself as a "universal citizen of humanity" is from Vejas Gabriel Liulevicius's *History's Greatest Voyages of Exploration*, "Lecture 15: Alexander von Humboldt: Explorer Genius" (Chantilly, VA: The Great Courses, 2015), although this was a common self-description at the time, as discussed in Anthony Pagden's *The Enlightenment: And Why It Still Matters* (New York: Random House, 2013). The brief summary of Ida Pfeiffer's travels is based on Liulevicius's "Lecture 18: Ida Pfeiffer: Victorian Extreme Traveler."

CHAPTER 16: LANDS OF ICE AND SNOW

The description of Sir John Franklin's expedition in this chapter owes much to Owen Beattie and John Geiger's excellent book *Frozen in Time: The Fate of the Franklin Expedition* (London: Bloomsbury, 1987) and to a lesser extent Gillian Hutchinson's *Sir John Franklin's Erebus and Terror Expedition: Lost and Found* (London: Bloomsbury, 2017). The quote "I beg your pardon? I'm only fifty-nine," comes from Vejas Gabriel Liulevicius's summary in *History's Greatest Voyages of Exploration*, "Lecture 17: Sir John Franklin's Epic Disaster" (Chantilly, VA: The Great Courses, 2015).

For a gripping description of the arctic voyage of *Fram*, Fridtjof Nansen's personal account *Farthest North: The Incredible Three-Year Voyage to the Frozen Latitudes of the North*, which was originally published in 1897, is a great source (abridged edition; New York: Modern Library, 1999).

For further reading on Nobile's arctic airship expedition and the resulting crash and then rescue, see Wilbur Cross, *Disaster at the Pole: The Tragedy of the Airship* Italia *and the 1928 Nobile Expedition to the North Pole* (Lanham, MD: Lyons Press, 2000).

The description of the race for the South Pole in this chapter draws on Stephen R. Brown's *The Last Viking: The Life of Roald Amundsen* (Boston: Da Capo, 2012) and Roland Huntford's *Scott and Amundsen: The Last Place on Earth* (New York: Abacus, 2012). But perhaps no book on Antarctic exploration is so compelling as Alfred Lansing's epic recital of Shackleton's struggle for survival in *Endurance: Shackleton's Incredible Voyage*, second edition (New York: Carroll & Graf, 1999).

CHAPTER 17: TO THE SKY

The summary of Amelia Earhart's life and aviation career in this chapter is based on Susan Butler's *East to the Dawn: The Life of Amelia Earhart* (Philadelphia: Da Capo, 2009) and Donald M. Goldstein and Katherine V. Dillon's *Amelia: The Centennial Biography of an Aviation Pioneer* (London: Brassey's, 1997).

Details on the search for Earhart and Noonan are mainly drawn from Randall Brink's *Lost Star: The Search for Amelia Earhart* (New York: W. W. Norton, 1994).

The International Group for Historic Aircraft Recovery has compiled evidence for various competing theories of what may have happened to Earhart and *Electra*. A summary of hypotheses and supporting evidence can be found at "The Post-Loss Radio Signals," TIGHAR.org (https://tighar .org/Projects/Earhart/Overview/AEhypothesis.html).

CHAPTER 18: THE SPACE RACE

This chapter's musings on how in engineering "know-how" often precedes "know-why," and also the trajectory of the Soviet and US Moon programs were greatly inspired by Professor David Mindell's course Engineering Apollo: The Moon Project as a Complex System, Massachusetts Institute of Technology, https://ocw.mit.edu/courses/science-technology-and-society/sts-471j-engineering-apollo-the-moon-project-as-a-complex-system-spring-2007/ (attended in 2006).

Other references consulted include Michael J. Neufeld's *Von Braun: Dreamer of Space, Engineer of War* (New York: Vintage Books, 2008), Annie Jacobsen's *Operation Paperclip: The Secret Intelligence Program That Brought Nazi Scientists to America* (New York: Little, Brown, 2014), and Andrew Chaikin's *A Man on the Moon: The Voyages of the Apollo Astronauts* (New York: Penguin Books, 2007).

CHAPTER 19: SEEING THROUGH ROBOTIC EYES

For further reading on the robotic exploration of the solar system, see Chris Impey and Holly Henry's excellent book *Dreams of Other Worlds: The Amazing Story of Unmanned Space Exploration* (Princeton, NJ: Princeton University Press, 2013). Another good general reference on the topic is *Worlds Beyond: The Thrill of Planetary Exploration as told by Leading Experts*, edited by S. Alan Stern (Cambridge: Cambridge University Press, 2003).

Alan Stern and David Grinspoon's *Chasing New Horizons: Inside the Epic First Mission to Pluto* (New York: Picador, 2018) provides a fascinating account on how robotic space exploration missions are organized, designed, and carried out (in particular the title mission to Pluto). I would also be remiss if I didn't mention Carl Sagan's *Pale Blue Dot: A Vision of the Human Future in Space* (New York: Random House, 1994), which convincingly argues why space exploration is well worth the cost, and also discusses planning for Voyagers 1 and 2 "grand tour" of the solar system.

For further reading on the Viking experiments, see Paul Chambers's *Life on Mars: The Complete Story* (London: Blandford, 1999); G. V. Levin and P. A. Straat, "Viking Labeled Release Biology Experiment: Interim Results," *Science* 194, no. 4271 (1976): 1322–29; and Rafael Navarro-González et al., "Reanalysis of the Viking Results Suggests Perchlorate and Organics at Midlatitudes on Mars," *Journal of Geophysical Research*, 115, E12 (2010), https://doi.org/10.1029/2010JE003599.

NOTES ON SOURCES

PART IV: BECOMING *STAR TREK*

CHAPTER 20: INTO THE FUTURE

Much of the discussion on artificial intelligence in this chapter was inspired by the fantastic documentary film *Do You Trust This Computer?*, directed by Chris Paine, http://doyoutrustthiscomputer.org/ (Papercut Films, 2018). Ray Kurzweil's *The Singularity Is Near: When Humans Transcend Biology* (New York: Penguin Books, 2006) was also consulted.

Additional musings on artificial intelligence and the human future draw from Michio Kaku's *The Future of Humanity: Terraforming Mars, Interstellar Travel, Immortality, and Our Destiny Beyond Earth* (New York: Doubleday, 2018), Peter Diamandis's *Abundance: The Future Is Better Than You Think* (New York: Free Press, 2012), and Carl Sagan's masterpiece works *Pale Blue Dot: A Vision of the Human Future in Space* (New York: Random House, 1994) and *Cosmos* (New York: Random House, 2002).

CHAPTER 21: THE ROAD TO MARS

Musings on NASA's trajectory early in this chapter were greatly inspired by Aircraft Systems Engineering (with a focus on the Space Shuttle), a course taught by Professor Jeffrey Hoffman and Aaron Cohen (former director of NASA's Johnson Space Center), Massachusetts Institute of Technology, https://ocw.mit.edu/courses/aeronautics-and-astronautics/16-885j-aircraft-systems-engineering-fall-2005/ (attended in 2005).

The estimate of just over a billion dollars per Saturn V flight is referenced in Roger E. Bilstein's *Stages to Saturn: A Technological History of the Apollo/Saturn Launch Vehicles* (Washington, DC: NASA History Office, 1997). The cost estimate of $450 million for a shuttle flight comes from Pat Duggins's *Final Countdown: NASA and the End of the Space Shuttle Program* (Gainesville: University Press of Florida, 2009).

The discussion of Mars as the best destination for humanity is heavily inspired by Robert Zubrin, who originally kindled my interest in the Red Planet with his fantastic book *The Case for Mars: The Plan to Settle the Red Planet and Why We Must* (New York: Free Press, 1996). Also consulted were Zubrin's *How to Live on Mars: A Trusty Guidebook for Surviving and Thriving on the Red Planet* (New York: Three Rivers Press, 2008) and *Mars Direct: Space Exploration, the Red Planet, and the Human Future* (New York: Jeremy P. Tarcher/Penguin, 2013).

CHAPTER 22: BECOMING SPACEFARING

Robert Zubrin's *The Case for Mars: The Plan to Settle the Red Planet and Why We Must* (New York: Free Press, 1996) and Peter Diamandis's *Abundance: The Future Is Better Than You Think* (New York: Free Press, 2012) inspired some of the projections in this chapter, in particular regarding energy production and resource extraction. Zubrin also presents some ideas for traveling beyond Mars in his chapter "Colonizing the Outer Solar System," in *Islands in the Sky: Bold New Ideas for Colonizing Space*, edited by Stanley Schmidt and Robert Zubrin (New York: John Wiley, 1996), 85–94.

Some specific thoughts on what a spacefaring future would be like (in particular, details related to establishing a human presence on Titan) owe their genesis to a reading of Charles Wohlfort and Amanda R. Hendrix's *Beyond Earth: Our Path to a New Home in the Planets* (New York: Pantheon, 2016). The idea of human settlement of Mercury as described in this chapter was proposed in Arthur C. Clarke's masterpiece *Rendezvous with Rama* (Orlando, FL: Harcourt Brace, 1973)—an example of science fiction inspiring science?

Chris Impey's *Beyond: Our Future in Space* (New York: W. W. Norton, 2016) was also consulted for this chapter, and the section on space food production was no doubt inspired in some way by Mary Roach's hilarious book *Packing for Mars: The Curious Science of Life in the Void* (New York: W. W. Norton, 2011).

CHAPTER 23: GOING INTERSTELLAR

Several authors were influential in the writing of this chapter. Chris Impey's *Beyond: Our Future in Space* (New York: W. W. Norton, 2016) gives a summary of certain forms of advanced propulsion systems, in particular the Alcubierre drive (originally proposed in Miguel Alcubierre, "The Warp Drive: Hyper-Fast Travel within General Relativity," *Classical and Quantum Gravity* 11, no. 5 [1994], https://doi.org/10.1088/0264-9381/11/5/001). Good overviews are also provided in I. A. Crawford, "The Astronomical, Astrobiological, and Planetary Science Case for Interstellar Spaceflight," *Journal of the British Interplanetary Society* 62 (2009): 415–21; Crawford, "Some Thoughts on the Implications of Faster-Than-Light Interstellar Space Travel," *Quarterly Journal of the Royal Astronomical Society* 36, no. 3 (1995): 205–18; and Geoffrey Landis, "The Ultimate Exploration: A Review of Propulsion Concepts for Interstellar Flight," *Interstellar Travel and Multi-*

Generation Space Ships, edited by Yoji Kondo et al. (Burlington, Canada: Apogee Books, 2003).

The concept of using nuclear bombs to power spacecraft was, to my knowledge, first proposed in C. J. Everett and S. M. Ulam, "On a Method of Propulsion of Projectiles by Means of External Nuclear Explosions," *Los Alamos Scientific Laboratory* (August 1955), reprinted in S. M. Ulam, *Analogies Between Analogies: The Mathematical Reports of S.M. Ulam and His Los Alamos Collaborators* (Berkeley and Los Angeles: University of California Press, 1999), http://ark.cdlib.org/ark:/13030/ft9g50091s/.

For further reading on laser-powered space travel, see Geoffrey Landis, "Laser-Powered Interstellar Probe," presented at the Conference on Practical Robotic Interstellar Flight, New York University, New York, August 29 through September 1, 1994. For the feasibility of generation ships, see Adam Crowl et al., "Embryo Space Colonisation to Overcome the Interstellar Time Distance Bottleneck," *Journal of the British Interplanetary Society* 65 (2012): 283–85.

Michio Kaku's *The Future of Humanity: Terraforming Mars, Interstellar Travel, Immortality, and Our Destiny Beyond Earth* (New York: Doubleday, 2018) and Robert Zubrin's *Entering Space: Creating a Spacefaring Civilization* (New York: Penguin Putnam, 1999) were also consulted for this chapter.

CHAPTER 24: LIFE ON OTHER WORLDS

For a thorough review of what we know about the prospect for life on other worlds, see *The Handbook of Astrobiology*, edited by Vera M. Kolb (Boca Raton, FL: CRC Press, 2019). Chris McKay does a good job reviewing what we know about the necessary conditions for life in "Requirements and Limits for Life in the Context of Exoplanets," *Proceedings of the National Academy of Sciences* 111 no. 35 (2014): 12628–33.

For background reading on panspermia, see Fred Hoyle and Chandra Wickramasinghe's *Evolution from Space: A Theory of Cosmic Creationism* (New York: Simon & Schuster, 1982) and Chandra Wickramasinghe, "Bacterial Morphologies Supporting Cometary Panspermia: A Reappraisal," *International Journal of Astrobiology* 10, no. 1 (2011): 25–30. For more on the possibility of RNA bridging the gap until the advent of DNA, see Marc Neveu et al., "The 'Strong' RNA World Hypothesis: Fifty Years Old," *Astrobiology* 13, no. 4 (2013): 391–403.

Discussion on extremophiles, evolution, and the Burgess Shale in this chapter was inspired by Bill Bryson's *A Short History of Nearly Everything* (New York: Broadway Books, 2003), Michael Land and Russell Fernald's "The Evolution of Eyes," *Annual Review of Neuroscience* 15 (1992): 1–29, and, of course, Richard Dawkins's fantastic books *The Greatest Show on Earth: Evidence for Evolution* (New York: Free Press, 2009) and *The Blind Watchmaker: Why the Evidence of Evolution Reveals a Universe without Design* (New York: W. W. Norton, 1996).

CHAPTER 25: THE ULTIMATE DESTINATION

For further reading on the Kardashev scale, see Nikolai S. Kardashev, "On the Inevitability and the Possible Structures of Supercivilizations," *The Search for Extraterrestrial Life: Recent Developments, Proceedings of the 112th Symposium of the International Astronomical Union Held at Boston University, Boston, Mass., USA, June 18–21, 1984*, edited by M. D. Papagiannis (Dordrecht, Netherlands: D. Reidel, 1984), 497–501.

The assertion that almost a quarter of Americans can't tell the difference between astrology and astronomy is from Elizabeth Palermo, "1 in 5 Americans Confuse Astrology and Astronomy," *LiveScience* (www.livescience.com/52135-american-science-knowledge-poll.html); it was 22 percent in the poll.

General intellectual fodder for this chapter was provided by Ray Kurzweil's *The Singularity Is Near: When Humans Transcend Biology* (New York: Penguin Books, 2006), Michio Kaku's *The Future of Humanity: Terraforming Mars, Interstellar Travel, Immortality, and Our Destiny Beyond Earth* (New York: Doubleday, 2018), and Michael Shermer's *Heavens on Earth: The Scientific Search for the Afterlife, Immortality, and Utopia* (New York: Henry Holt, 2018).

Index